Managing Coastal and Inland Waters

Managing Coastal and Inland Waters

Pre-existing Aquatic Management Systems in Southeast Asia

Kenneth Ruddle and Arif Satria
Editors

Ashiya City, Hyogo, Japan
Bogor, Jawa Barat, Indonesia

Springer

Editors
Prof. Kenneth Ruddle
7-22-511, Asahigaoka-cho
Ashiya City, Hyogo 659-0012
Japan
mb5k-rddl@asahi-net.or.jp

Dr. Arif Satria
Jalan Bambu Raya 30
Taman Yasmin Sektor 7
Bogor
Jawa Barat
Indonesia
arifsatria@ipb.ac.id

ISBN 978-90-481-9554-1 e-ISBN 978-90-481-9555-8
DOI 10.1007/978-90-481-9555-8
Springer Dordrecht Heidelberg London New York

Library of Congress Control Number: 2010935584

Cover illustration top: Kenneth Ruddle, 1985
Photo text: Small coastal sailing craft for general cargo in Kalbut Village, East Java, Indonesia

Cover illustration bottom left: Kenneth Ruddle, 1995
Photo text: Woman fish buyer and fish seller butchering sharks at the main fish market in Danang City, Vietnam

Cover illustration bottom right: Kenneth Ruddle, 1995
Photo text: Landing the small pelagics catch in Le Hieu Hao Village, Nhatrang, Khanh Hoa Province, Vietnam

Printed on acid-free paper

Springer is part of Springer Science+Business Media (www.springer.com)

Contents

List of Figures

List of Photographs

List of Tables

Contributors

Dedi S. Adhuri received his Ph.D. in Anthropology from the Australian National University. He now works in Policy Economics and Social Sciences, at the WorldFish Center, Penang, Malaysia. He is also affiliated with the Maritime Study Group at the Research Center for Society and Culture, Indonesian Institute of Sciences (LIPI), in Jakarta. With both institutions, he has carried out intensive research on marine resource management and post-disaster coastal and livelihoods rehabilitation. E-mail: dediadhuri@hotmail.com

Ian G. Baird has a Ph.D. in Geography from The University of British Columbia, Vancouver, Canada. He is the co-founder and Executive Director of the Global Association for People and the Environment (GAPE), a Canadian non-government organization active in southern Laos. He has been working and living in Laos and Thailand for most of the last 23 years. He has also worked extensively in northeastern Cambodia. In Laos he has conducted extensive research and published widely on fish, fishers and fisheries, as well as various other nature-society-development issues. E-mail: ibaird@wisc.edu

Malasri Khumsri is a Senior Fisheries Biologist in the Department of Fisheries, Thailand. She received a M.Sc. degree in Aquaculture and Aquatic Resources management in 2002 and a Ph.D. in 2008 from Asian Institute of Technology (AIT). She also serves as a national coordinator of Thai Fisheries Management and Governance sub-component, Fisheries Program, Mekong River Commission (MRC). She has working experiences in fisheries resources management as a sustainable livelihoods and fisheries biologist in the Mekong Region. Her research interests are fisheries management, conflict resolution in natural resources management, institutional analysis, community-based management and sustainable livelihoods. E-mail: kmalasri@yahoo.com

Maria F. Mangahas is an Associate Professor in the Department of Anthropology, University of the Philippines, Diliman. She received her Ph.D. in Social Anthropology from the University of Cambridge. Dr. Mangahas has done field research in small-scale fishing communities in Batanes Province (Northern Philippines) and Samal Island, Davao (Southern Philippines), and has published or presented papers on indigenous coastal resource management, collective fishing technology, fishers' perceptions on fisheries, impoverishment and sharing, notions of 'luck' and leadership, and gear conflicts and social change. At present she is researching the phenomenon of 'digitized scandal' and the culture of media piracy in the Philippines. E-mail: maria.mangahas@up.edu.ph

Duy Thieu Nguyen hails from Ha Tinh Province, Vietnam. In 1976 he received a B.A. in Ethnology from the University of Hanoi, and in 1994 a Ph.D. in Ethnology from the Vietnam Academy of Social Sciences. Since 1976 he has been a Senior Research Fellow at the Institute for Southeast Asian Studies, Vietnam Academy of Social Sciences. He also gives lectures in ethnology in several universities in Vietnam, and is Vice Director of the Vietnam Museum of Ethnology. At present, Dr. Thieu is researching the life of fisheries communities in Vietnam, and also of the Thai-Kadai speaking ethnic groups in Vietnam and in Lao PDR. He is also studying Overseas Vietnamese and anthropological museology. E-mail: thieund_dna@yahoo.com

Kenneth Ruddle received a Ph.D. from the University of California, in 1970. At present he is a Professor in the School of Policy Studies of Kwansei Gakuin University, in Japan. He specializes in the human ecology of tropical regions, and has done extensive fieldwork in South America and Africa, in addition to East Asia, his main area of focus. Dr. Ruddle's main research interest is on pre-existing systems of tropical aquatic resources management and their local knowledge base. He was awarded a Guggenheim Fellowship in 1980–1881, and is a Fellow of the World Academy of Art and Science. E-mail: mb5k-rddl@asahi-net.or.jp

Arif Satria is Dean, Faculty of Human Ecology, Bogor Agricultural University, Indonesia. He obtained a M.S. in Rural Sociology from the Graduate School of Bogor Agricultural University, and his doctorate from the Department of Marine Production and Environmental Studies, Kagoshima University, Japan. His research interests include fisheries sociology, political ecology, marine policy, and community-based fisheries management. E-mail: arifsatria@gmail.com and arifsatria@ipb.ac.id

Chapter 1
An Introduction to Pre-existing Local Management Systems in Southeast Asia

Kenneth Ruddle and Arif Satria

Abstract Although known from colonial times, pre-existing systems of fisheries management in tropical nations have not usually been used as an alternative to introduced Western scientific approaches. During the colonial era non-Western models were disparaged openly, whereas nowadays commonly they are dismissively labeled as 'traditional' or 'special' cases. Often predicated on misguided theories, during the 1950s and 1960s a massive and experimental packaged transfer of social, economic, financial, educational, and legal systems, together with their underlying cultural values and aspirations regarded pre-existing economies, management systems, and often social and cultural systems as obstacles to modernization. Modernization provided the justification for foreign designers of fisheries management schemes to claim that pre-existing systems were either primitive or unsustainable or often 'non-existent'. This was reinforced by a general ignorance of the tropics and prejudice on the part of scientists and educators, whose careers were enhanced by work in temperate regions. The generic 'design principles' and functioning of pre-existing systems is summarized, together with the status of knowledge on Indonesia, Laos, the Philippines, Thailand, and Vietnam.

Keywords Design principles • Geographical distribution • Management functions • Tropics

1.1 Introduction

From the late-1970s and early-1980s pre-existing (commonly termed 'traditional', 'customary' or '*de facto*') rights-based fisheries management became an important research topic (Cordell 1977; Johannes 1978; Ruddle and Akimichi 1984; Ruddle

K. Ruddle (✉)
Asahigaoka-cho 7-22-511, Ashiya City, Hyogo-ken 659-0012, Japan
e-mail: mb5k-rddl@asahi-net.or.jp

A. Satria
Jalan Bambu Raya 30, Taman Yasmin Sektor 7, Bogor, Jawa Barat, Indonesia
e-mail: arifsatria@ipb.ac.id; arifsatria@gmail.com

K. Ruddle and A. Satria (eds.), *Managing Coastal and Inland Waters: Pre-existing Aquatic Management Systems in Southeast Asia*, DOI 10.1007/978-90-481-9555-8_1,

and Johannes 1985). Although they were clearly a well-understood problem for early colonial administrations in many locations (Ruddle 1995, 2007b), only in the last 30 years has the modern usefulness of pre-existing rights been acknowledged as an important factor in fisheries management (see, for example, Fa'asili and Kelokolo 1999; Hickey 2006; Johannes 1977, 1981, 1994, 2002, 2003; Johannes and Hickey 2004; Ruddle 1998a; Tiraa 2006; Veitayaki 2001). It has now been conclusively demonstrated, as in Samoa (Fa'asili and Kelokolo 1999), Solomon Islands (Aswani and Hamilton 2004), and Vanuatu (Johannes 1998; Johannes and Hickey 2004), among other places, that pre-existing rights may be used to design and exercise the rights of management and exclusion, which would work as an incentive in collective action for the improvement of fisheries use and management.

As is well known from copious subsequent research, such systems are based on common property rights concepts. One of their advantages for small-scale fishers is that risk and uncertainty about resources and social organization is reduced. Risk and ill-affordable wasted effort is greatly reduced because fishing behavior is based on local knowledge of resources and resources are protected by controlling the access of outsiders; and social risk is reduced because cooperation and reciprocity, among other values, are emphasized and reinforced (Ruddle 1989a).

Although understanding property rights systems is of course basic to understanding the local management of resources, paradoxically, the definition, usage and general misunderstanding of the various types of 'property' has often impeded advances in theoretical thinking. Worse, it has been perversely detrimental when applied in the context of development (Bromley 1992).

Since the early-1950s examination of common property resources and collective goods by political economists yielded the policy prescription that, to achieve their potential social benefits, collective goods must be administered by a centralized authority, a pernicious and erroneous conclusion that continues to permeate contemporary policy prescriptions. Worse, when units of a central or local government either fail to perform or are deemed incapable of performing, privatization, often in the form of ITQs in fisheries, for example, is recommended as an unconvincing panacea (Ostrom 1998:6). Although government centralization ideas remain pervasive (Ostrom 1990, 1998), the management paradigm shifted in the 1980s and 1990s from external coercion to public participation, community-based management of 'collective goods', and co-management, changes which occurred within the context of comprehensive changes in approaches to national development and assistance (Ruddle 2007a).

Ostrom (1990) challenged both scholars and development practitioners with the essential need to "map the terrain" (Ostrom 1990:214) for a family of models, and not just one particular model, in order to improve practical outcomes, for in that direction alone is the escape from the "trap of omniscience." In a criticism of reliance on narrowly-conceived models as the foundation for policy analysis, Ostrom (1990: 215) trenchantly writes that "[w]ith the false confidence of presumed omniscience, scholars feel perfectly comfortable in addressing proposals to government that are conceived in their models as omnicompetent powers able to rectify the imperfections that exist in all field settings." In these models pre-existing local systems of rules for property management are either not recognized or willfully discarded (Ruddle 2007b; Ruddle and Hickey 2008). Worse, the models reinforce the role of government, often while masquerading as those aimed at decentralization! Not only does this toss aside perfectly viable management systems, it also adds to the tasks of governments that are either not competent to handle new challenges, or already absorbed with other tasks often erroneously perceived as more important.

1.2 Why Pre-existing Systems are Overlooked

Pre-existing fisheries management systems have been overlooked for several reasons (Fig. 1.1). The underlying reason for the failure to consider them as alternative models for management of tropical nearshore fisheries stems from a continuing legacy of colonialism and cultural imperialism demonstrated in donor and development agency behavior. It prevents a fuller consideration of the basic principles and operational designs of many pre-existing non-Western systems of proven viability, and instead continues to rely on unproved models and approaches designed largely by Western fisheries biologists, social scientists and policymakers. The earlier Western approaches of colonialism, technical assistance and financial aid have been repackaged as development aid conditionalities, technical expertise and the training

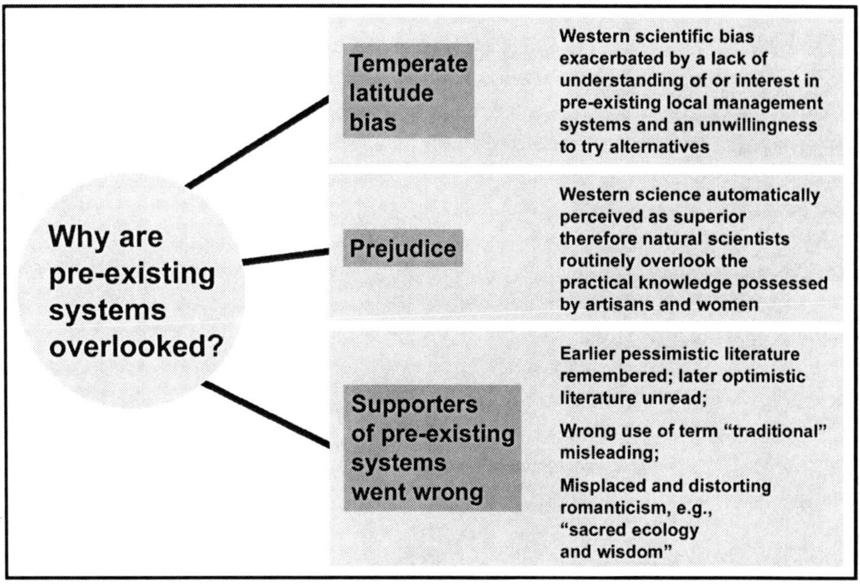

Fig. 1.1 Why pre-existing systems are overlooked

of local professionals. And whereas non-Western models were formerly disparaged openly, these days disparagement is more subtle; it often takes the form of labeling them as 'traditional' or 'special cases', and then dismissing them with no further discussion, while the 'serious' examination then turns to the Western models, with which authors are familiar (Ruddle and Hickey 2008).

That is compounded by the continuing historical legacy of colonial interventions. In the Asia-Pacific Region the colonial era had a major and lasting impact on pre-existing systems for managing nearshore fisheries. Its main impacts include undermining or displacing pre-existing tenure systems in conjunction with an added legal complexity, with the Western-based State law at odds with local customary law. In general, Western-trained lawyers believe that customary law is invalid for upholding legal claims and inferior to the Western legal tradition. This has been a major 'external' factor that either deliberately or by default undermined customary law and community resource rights (Ruddle 1994a, b, c, 1995, 2007a).

Further, the 1950s and 1960s witnessed the massive but experimental transfer of technology and capital from the rich to the poor nations. Social scientists concocted a Modernization Theory to justify the packaged transfer of social, economic, financial, educational, and legal systems, together with their underlying cultural values and aspirations. Such a transfer necessitated the prior destruction, side-lining or ignoring of pre-existing economies, management systems, and often social and cultural systems, which were regarded as obstacles to modernization (e.g., Seibel 1994). It was of little importance that such a theory of modernization was often predicated on misguided sub-theories. One was Garret Hardin's influential yet erroneous thesis of the 'tragedy of the commons' (Hardin 1968) which, gave justification for foreign designers of fisheries management schemes to claim that either pre-existing systems were unsustainable or that none existed.

The situation is exacerbated because tropical nearshore fisheries development projects are characterized by a Western scientific bias, worsened by a general lack of interest in or willingness to understand pre-existing local systems. Conservation of fish stocks became the main goal of development assistance, based on transplanted Western fisheries management models, with fisheries policy and management based on a conventional Temperate Zone bioeconomic model.

Most fisheries biologists and the social scientists often have only limited experience in the tropical milieux. Not surprisingly, therefore, they commonly fail to appreciate differences between the temperate zone industrial fisheries, with which they are familiar from their own training and research, and tropical nearshore fisheries. This means that erroneous interpretations are passed to donors and assistance personnel. Usually, it is not widely appreciated that in tropical nearshore fisheries (i) fishing is limited geographically to nearshore areas (ii) that are defined socially, and that (iii) fishing communities are numerous and dispersed geographically. It is also not commonly appreciated that (iv) tropical nearshore fisheries are biologically and technically complex compared with temperate areas, and are typically far more varied in terms of catch composition or areas fished and gear types employed. Hence they are of unfamiliar complexity to temperate region scientists and planners, who typically deal with single-species fisheries. That (v) employment options are limited and alternative jobs scarcity all-pervasive is often disregarded.(vi) Geographical and social territoriality is widespread, which, in addition to its positive aspects in terms of resource management, limits the mobility of small-scale fishers geographically and socially, and prevents access to fishing communities by outsiders. With regard to (viii) economic rent extraction, it needs to be appreciated that those various factors combine to create market imperfections such that nearshore fishers in many tropical regions receive less than the free-market price for their catch, yet pay excessively for inputs and usuriously for loans. These are the principal ways in which rents are extracted. They are also extracted by the requirement to share catches in small, traditional communities and among kin, as well as by other customary practises, such as ritual performance and donation (Ruddle 2007b; Ruddle and Hickey 2008).

Many of those difficulties could be overcome were it not for the persistence of an extremely negative connotation associated with the term "tropics" among fisheries scientists based in the temperate latitudes. Pauly (1994) summarized the prevailing attitude in an insightful essay inspired by a peer review which in its entirety read "Rubbish, may apply in the tropics – but not here".

There is little doubt that an elitist bias virtually deifies objective Western science and regards other knowledge systems as illegitimate, and those who challenge conventional theories and formal models are belittled. Such deeply embedded attitudes inhibit unconventional projects and research, and innovation is dissuaded when only empirical, quantitative methodologies are acceptable. This results in a standardized technological transfer being promoted by the structure of research institutions and professions. Indeed, Johannes (1981) contended that the crux of the issue that handicaps the development of nearshore tropical fisheries is the lack of integration of knowledge with elitist natural scientists routinely overlooking the practical knowledge possessed by artisans.

The historical roots of this prejudice are deep. One of the massive if insidious impacts of both historical and contemporary globalization has been the imposition of standard Western systems of resource management. In every respect this is the cultural equivalent of a major reduction in biodiversity. Coastal communities throughout the tropics and elsewhere (like the New Zealand Maori, for example [Ruddle 1995, 2007a]) experienced this early in the colonial era, when many communities were wrongly deprived of their traditional rights to fisheries and other resources. In some cases these have only recently been restored to them (Ruddle 2007a).

The most pernicious impacts of this conventional and long-applied Western model derive from the modern assumption of the lack of prior local institutional arrangements among fishers to govern a fishery, and that fisheries are unregulated by local collective action. The bioeconomic management model therefore argues that to manage stock externalities institutional arrangements must be imposed on local fishing communities by some outside level of government. Such schemes are based on the assumption that the institutional context of the fishery is one of open access. This is simply not true for vast tracts of the world's nearshore waters, particularly in tropical regions.

There are several reasons why those of us who have long emphasized the practical importance of considering pre-existing management systems are also partly responsible for their having been overlooked. One important reason was the pessimism expressed in one of the earliest articles on pre-existing systems (Johannes 1978). In the mid-1970s pre-existing systems of community-based marine resources management were everywhere in decline, the victims of westernization. Despite their functional elegance, Johannes was naturally pessimistic about their future. Unfortunately, many of those who read this early article based their unwavering opinions on it. Probably many did not bother to keep up with the related literature over the next 25 years, and more than likely they failed to read Johannes' (2002) follow-up article. Belying Johannes' pessimism of a quarter century earlier, an amazing transformation had occurred in the ensuing 25 years, particularly in Vanuatu and Samoa, where new fisheries management designs have been based largely on the pre-existing systems. We all felt vindicated.

The second reason has undoubtedly been the use from the very beginning of the notoriously imprecise term 'traditional', as in 'traditional management' and 'traditional (ecological) knowledge' (Davis and Ruddle 2010). This has probably not presented pre-existing systems in either an accurate or favorable light. Worse, its use enables proponents of Western management models to claim that if something is 'traditional' ipso facto it is unsuited to modern conditions. In particular, it provides a perfectly tailored excuse for donors with different agendas, like participatory democracy cloaked in a co-management design, to claim, for instance, that chiefly authority of 'traditional management systems' is undemocratic and therefore antithetical to modernization. Further, some tropical societies may see the term 'traditional' as pejorative and synonymous with 'backward', which might incline them to accept a Western management model as part of a development assistance package.

Third is that the uncritical acceptance and romanticizing of 'traditional' ecological knowledge, inflated claims about its environmental wisdom without determination of its validity, and selectively using facts to fit pre-conceived cases, have provoked a backlash. Particularly regrettable has been the conflating of an imputed sacredness with profound ecological wisdom, or the use of such phrases and terms as 'sacredness of ecological systems' or 'sacred ecology' (Berkes 1999; Durning 1992; Johannes 2003; Ruddle 2007c; Davis and Ruddle 2010).

1.3 Geographical Distribution of Pre-existing Fisheries Management Systems

Pre-existing systems of marine resource management are or were utilized widely throughout the Asia-Pacific Region to manage coastal fisheries. Such systems are particularly widespread in the Pacific Islands (Ruddle 1996a). Existing examples in Asia have been documented over a wide yet discontinuous geographical range, extending from Japan to Sri Lanka. Further, time-honored fisheries management systems are widespread throughout the world; they also occur in the Caribbean,

South America, Africa, and the Middle East. They are not restricted to developing countries. Similar systems are used by both aboriginal populations and communities of European ancestry in North America, Australia and New Zealand, as well as in several countries of Europe, and in Japan (Ruddle 1994c).

More recently, inland water resource management systems have been examined in continental Southeast Asia, especially in Laos (Baird 2006a, b; Baird et al. 2003; Tubtim and Hirsch 2005) and Thailand (Kuaycharoen 2002, Khumsri et al. 2009). Elsewhere, research on lacustrine rights systems has been done in floodplain lakes in the Brazilian Amazon (Mcgrath et al 1993), Lake Biwa, Japan (Kada 1984) and Lake Titicaca (Levieil 1987), for riverine fisheries in Brazil (Castro de and Begossi 1995; Silvano and Begossi 1998, 2001), and in several locations in Africa, such as Lake Chad and adjacent areas of West Africa (Sand 1970; Sarch 1994; Neiland et al. 1994).

1.4 Status of Information on Pre-existing Systems in Southeast Asia (Fig. 1.2)

1.4.1 Indonesia

Pre-existing fisheries management systems were formerly widespread in Indonesia (Ruddle 1994c). Although they were noted and minimally described in parts of Sumatra, there is little information for Western Indonesia (Polunin 1984). In North Sulawesi a system known as *seke* was described by Wahyono (2000). In South Sulawesi the *rompong*, an old established form of marine tenure that originated in the Bugis community at Makassar, was described by Zerner (1989a, b, c, 1991a) and by Saad (in Satria et al. 2002). On Ambo Island, in the Balabalakang Islands of the Makassar Strait, Kalimantan, Zerner (n.d.) described an indigenous royalties system employed to regulate the harvest of resource-rich areas within about 3 km of the high water mark. Satria (2007) examined the *awig–awig*, a pre-existing system in Lombok, West Nusa Tenggara Province. Pre-existing marine tenure in Irian Jaya was described in Wahyono (2000).

In contrast there are more comprehensive descriptions for pre-existing fisheries management systems in the central and southeastern Maluku Islands, where they remain more widespread than elsewhere in Indonesia, except possibly for Irian Jaya Province.[1] The *sasi laut* of Maluku has been the focus of discourse on pre-existing marine management systems in Indonesia since the 1980s, when it was examined widely by NGOs, research centers and legal scholars. The resultant publications (e.g., Anon 1991; Bailey and Zerner 1992; Kissya 1995; Naamin and Badrudin 1992; Pusdi-PSL Unpatti 1995; Zerner 1991a, b, c, d) contributed greatly to the understanding of pre-existing marine tenure, although they sometimes exaggerated its merits (e.g., Lokollo 1988).

[1]Which belongs to the Melanesian culture realm, rather than to Southeast Asia.

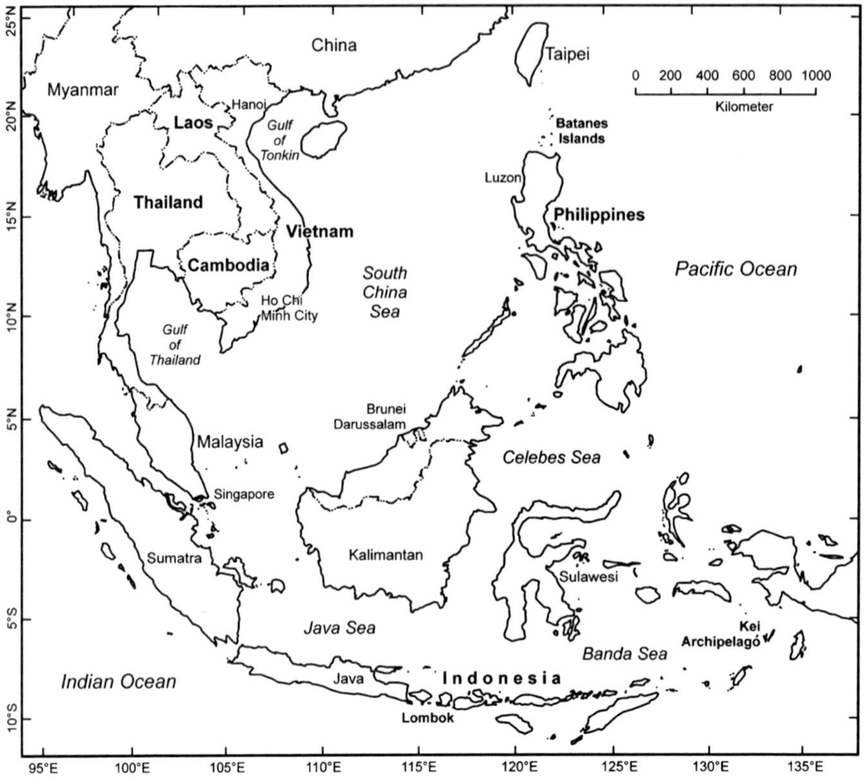

Fig. 1.2 Locations in Southeast Asia

These earlier works were later criticized as misleading because they failed to consider the historical and sociopolitical context of sasi (e.g., Pannell 1997). Seen from that perspective, considerable changes were revealed in sasi over four centuries, from an early ritual protection of communal resources, through a government-regulated agro-ecological control of private and common resources, to a largely commercialized and privatized means of theft prevention (Benda-Beckmann et al. 1992). In that way it was demonstrated that both local and outside elites had played a major role in changing sasi, such that by the late-colonial era, for example, the ratification of sasi rules was initiated by local traditional elites, in collaboration with local Dutch officials, to meet the economic and political interests of both parties. Elite control had also occurred during the 1960s, with respect to Topshell (*Trochus niloticus*) in Nolloth village on Saparua Island (Zerner 1991c), for example.

Such historical and contemporary analyses challenge the conservation and equity aspects earlier attributed to sasi. When analyzed in socio-political context, it is evident that local traditional leaders, NGOs and scholars have been actively engaged in the political process of 'greening' sasi, with the aim of empowering marginalized local people. Of this Zerner (1994) observed that the political context of the emergence of green sasi included both an increasing environmental awareness and also the resistance of local elites and NGOs to growing resource control by the central government and fishing industry. The early discourse on sasi emerged in response to the failure of centralized marine resource management, since sasi was seen as a better alternative method of marine resource management, that would ensure a more equal sharing of resources, prevent user conflict, and contribute to resource sustainability. As the discourse has developed, however, these attributes became questioned. For example, historical accounts found that conservationist motives had not driven the practice of sasi. Rather, economic and political interests were dominant in shaping it in both historical and contemporary settings.

The discourse on sasi has not adequately addressed the politically and economically important issue of resource ownership. Further, it cannot explain the increasing tension over marine resource management, perhaps because it tends to take the issue of marine ownership for granted and, as represented by such concepts as *petuanan laut* or *meti*, considered to be not problematical. Looking at the practice of sasi in Nolloth, for example, after the village head applied it to *Trochus niloticus*, the issue of rights over territory and the resource itself came into question. The effects of different attitudes toward issues of sea ownership are evident in many of the cases examined by Adhuri (2002). For example, in relation to marine resource management, Kei Islanders talk more about petuanan laut (sea territory) or meti (coastal water) than sasi. For them communal sea ownership (*hak ulayat laut*) is more important than sasi. In fact, the practice of sasi in Sather, Tutrean and Hollat villages in eastern Kei Besar Island cannot be carried out properly, owing to conflict over sea territory between and within them.

Studies on the contemporary practice provide further insights into the local realities of sasi. Pannell (1997: 297) noted that

> [T]he practices referred to and associated with sasi in the marine environment of Luang [south-eastern Maluku] minimally involve the interest and actions of residents of this island, the commercial machinations of regional traders and internationals exporters, the fashions and fads of distance consumers, the compliance and blessing of the Church and its agents, as well as the endorsement of village representatives of local government institutions and the support of government personnel from other jurisdictions. In addition, let us not forget those fishermen who, through their non-sanctioned exploitation of local marine resources, contribute to the social delimitation of the efficacy of invoking sasi.

Having noted such varied involvement in the practice of sasi, Pannell (1997: 296) made the unsurprising observation that it might mean different things to different agencies with different interests.

A doctoral dissertation by Adhuri (2002), based on field research in the Kei Islands from February 1996 to March 1997, not unexpectedly concluded that the practice of pre-existing marine tenure is embedded in the social world of the community, and therefore a people's perception and practice of a given system is always connected to the social structure of the community. That commonsense approach elucidates how systems adapt to and deal with problems associated with modernization and market development.

Adhuri examined the characteristics of communal marine tenure (*hak ulayat laut*), the legal status of pre-existing marine tenure and the effect of legal status on it, and the impact of the market economy on the practice of communal property rights. He argued that the practice of pre-existing marine tenure in the Kei Islands is far more complex than some of the early discourse suggested, particularly in relation to the issue of territorial rights, since both the rights of *hak makan* ('right to eat', use right) and *hak milik* (right of ownership) are attached to claimed sea territory. His research suggests that only hak makan is distributed equally among members of the community, whereas hak milik is the exclusive right of only particular segments of it.

However, it is important to note that definition of the right-holding unit in relation to both rights is subject to ongoing and sometimes violent dispute, as are the boundaries of a particular sea territory. That the territorial and social units concerned are often disputed reflects that traditional marine tenure in the Kei Islands is linked integrally to the social structure of factionalized communities. In fact, control over sea territory is a symbol of precedence. When contestation over precedence arises within communities, control of marine resources automatically becomes one of the issues contested and various sources of legitimacy, both state and traditional, are drawn on to support a variety of claims. Adhuri's argument concerning the legal aspect of traditional marine tenure is that formal government acknowledgment of the practice does not guarantee that people would always use the pre-existing marine tenure institution to guide their behavior. This is because, at a practical level, power and interest are much more important than formal legal status. In the Kei Islands, whether or not the government formally acknowledges pre-existing marine tenure is debatable. But, when it comes to practice, it is not the law and regulations that define whether pre-existing marine tenure principles and procedures are referred to; rather it is the distribution of power that determines the choice, regardless of either formal laws or pre-existing principles. This pragmatic approach reflects that many Kei Islanders do not necessarily see modern state law as replacing pre-existing arrangements. Rather they see the process as being cumulative, whereby state laws provide new options for them in the pursuit of their resource claims.

Finally, Adhuri's research in the Kei Islands, suggests that the market economy does not always necessarily degrade pre-existing marine tenure. His evaluation of the impact of the international trade of live reef fish and frozen anchovy shows that people have strengthened pre-existing marine tenure to exclude others from their territory. In other words, people manipulate and revive traditions to secure access to the newly valuable resources.

1.4.2 Laos

Although not yet comprehensively studied, many pre-existing fisheries management systems remain in Laos, although increasingly they are being modified to meet current conditions (Claridge et al. 1997; Baird 2006a). There appear to be three broad groups of such systems: areas seasonally or permanently closed, gear limitations, and the protection of individual species (Claridge et al. 1997).

Systems in southern Laos, particularly in the Khone Falls area in southern Khong District, are now quite well understood, thanks largely to the work of Baird and his associates. Complex individual and family tenure systems were developed to apportion the limited good trap setting sites for particular fisheries (Roberts and Baird 1995). An example is the gill net site tenure system for the *Probarbus* fishery at Hang Khone and Hang Sadam (Baird 2006b). This is essentially a private tenure system over fishing sites, with claims based on first-comer rights, which, once established, become de facto private property in perpetuity. Sites are occasionally rented by the season, with ownership rights returning to the original claimant at the end of the agreed period.

Many different types of fish sanctuary exist throughout Laos under a wide range of social, cultural, geographical and ecological situations. The main types were reviewed by Baird (2006a), with a particular emphasis on those in Khong District, Champasak Province, and deep-water pools in the mainstream of the Mekong River. In Luang Prabang Province of northern Laos, 37% of the village representatives reported that their communities had nearby fish sanctuaries, mainly associated with deep-water pools in rivers, which locals believe are important fish breeding grounds, and often constitute important dry-season refuges. The situation is similar in Sayabouli Province (Baird 2006a). In a karstic area of Khammouane Province, central Laos, deep caves filled with water adjacent to the karsts often become the only areas with water during the dry season, when they function as fish refuges that are not fished. Streams flow again with the onset of the rainy season, and these pools are often the sources of the larger broodstock for local fisheries (Shoemaker et al. 2001). In many parts of Laos distant from large rivers, deep-water parts of enclosed natural wetlands function as fish sanctuaries and are governed by seasonal fishing restrictions during the wet monsoon, and opened for limited periods for fishing when water levels become critically low at the height of the dry season (Tubtim and Hirsch 2005). Fish sanctuaries are managed by local institutions that establish rules and mete out sanctions on violators. Authority is often vested in village elders or the village chief, who manage fish sanctuaries on behalf of their communities (Baird 2001, 2006a).

Management systems of small wetlands or backswamps are locally varied and change according to water levels; access is open during the rainy season when floodwaters obscure boundaries, and excusive rights return as clear boundary markers emerge again in the dry season. Most are governed by customary rules that often involve spiritual beliefs (Tubtim 2006).

Supernatural authority still plays an important role for many of the pre-existing fishery management systems (Tubtim 2006). For example, Baird (1999) reports that most of the limits on fishing and other aquatic resource harvesting activities in the Khong District are related to reducing risk from dangerous spirits (*phi*), crocodiles (*khe*), large sting-rays (*pa fa lai*) or serpents (*gneuak*), which placed specific deep-water in the Mekong River off limits. And village fish sanctuaries are often associated with animist beliefs. Areas consciously protected for religious reasons also can function as fish sanctuaries, although not specifically recognized as such (Baird 2006a).

1.4.3 Philippines

There is little documentation on pre-existing marine resource management systems in the Philippines. However, more detailed field research is likely to reveal their widespread existence, since they appear to have been commonplace historically.

During the Spanish administration, fisheries were for the welfare of the town (municipality) and were an open access resource, although private rights were leased to individuals, particularly for construction of fish corrals (Spoehr 1980). The U.S. administration, by Act 4003 Sect. 67 (1932), authorized municipalities to grant exclusive fishing rights to concessionaires within municipal waters via public auctions, principally with the intent to generate revenue for the municipality. Where this did not occur gear was licensed to derive revenue (Santos 1980). However, the Supreme Court ruled that the auction of exclusive rights pertained only to stationary weirs (fish corrals), oyster beds or fry collecting, and that municipalities could not exclude non-resident fishermen who obtained a permit from the Bureau of Fisheries and Aquatic Resources (Kalagayan 1991). Under the Fishery Decree of 1975 the licensing authority for municipal waters was given to the municipalities.

According to early Spanish chroniclers, systems of community-based coastal and riverine fisheries management existed in pre-Hispanic and early Colonial times, based on independent villages (*barangay*), around Manila and in the Tagalog Region of Luzon Island (Blair and Robertson 1903–1909). Around Manila, villages claimed areas of river and sea that were defensible against neighboring settlements (Francisco Colin 1663, cited in Blair and Robertson 1903–1909). In the Tagalog Region village fishing territories could be used by outsiders on payment of a fee (Juan de Plasencia 1589, cited in Blair and Robertson 1903–1909), and such territories could be bartered like other property (Francisco Combes 1667, cited in Blair and Robertson 1903–1909). However, as Spanish colonial rule intensified the barangay was eliminated as an administrative unit, and its village sea territory disappeared with it (Lopez 1985).

However, it is evident that the tradition did not disappear entirely. There are several examples.

In the Lingayen Gulf of Luzon Island there is a de facto system of access restriction associated with traditional types of fish aggregation device (*rama* and *radar*), since those who established them have the exclusive right to fish around them (Galvez 1991). Artificial reefs made of tires were introduced to the Agoo Municipality, in La Union Province, in 1981. Although fishermen from various villages assisted in their construction, those who were not members of the Agoo cooperative were prevented from fishing near them. Those who were allowed to work the artificial reefs had to pay a portion of their catch to the president of the cooperative.

At Quinlogan Village, Palawan Island, fishing was conducted historically under open access. However, with the arrival of migrants from the Visayan Islands, from 1960, concepts related to the national law of municipal fisheries were introduced. But, within the statutory law local management rules have been introduced to regulate beach seine operations. This was done in the 1980s to allocate equitably access to the prime site for catching shrimp fry. Hitherto, operations were based on first-comer's rights, but crowding made necessary some kind of allocation mechanism. At the start of the season beach seine operators meet to allocate turns to the prime site for all operators. After the assigned operator has set his net, all others may set theirs, on a first-comer basis (Veloro 1992).

In San Miguel Bay, in the Bicol Region of Luzon Island, rights to fixed gear sites have been traditionally allocated to individuals by informal village resource managers, *amoionadors* ('boundary setters'). These are mainly respected village elders well versed in the history of family claims to fishing sites. Their primary task is to regulate new entrants to minimize conflict with established gear (Cruz 1982, 1986). This has become formalized at the village level in more congested areas, where the amoionadors charge a fee for their services (Cruz 1982), whereas in other areas their role is still informal and traditional. Elsewhere, municipalities regulate details of gear placement to minimize conflict (Cantero-Pastrano 1955; Hart 1956).

Milkfish (*Chanos chanos*) fry concessions were examined during the 1980s (e.g., Smith 1981; Chong et al. 1982; Smith and Panayotou 1987). The municipality, as the resource owner, granted to the highest qualified bidder the exclusive right to gather milkfish fry from municipal waters for a period not exceeding five years. Sealed bids were submitted annually on a designated date. Such concession fees comprised an important part of the income of many municipalities. In a survey of 35 fry grounds, milkfish concession fees represented an average of 13% of municipal income (Smith 1981), and in the Western Visayan Province of Antique 21% of the income of the 15 municipalities was derived from such concessions. Several municipalities obtained almost half their income in this way (Smith and Panayotou 1987). In small coastal municipalities this income was used to pay the salaries of municipal officials and the allowances of the Municipal Council (Smith and Panayotou 1987). Fry collection was done by laborers who had to sell their catch to the concessionaire.

1.4.4 Thailand

Pre-existing fisheries management systems in Thailand have been studied only recently. Field research in the Lower Songkhram River Basin of Northeast Thailand has revealed a complex situation. At present, fisheries resources there are managed concurrently by local communities, based on pre-existing rights, and de jure by the Department of Fishery (DoF), according to the *Fisheries Law* of 1947. Further, according to the *Thai Civil and Commercial Law* of 1925, natural resources used in common, such as shores, streams and lakes, are State Property (RTG 1930). Concurrently, local communities recognize that individuals have ownership of fishing rights in such areas, and that they also have the right to exclude others from fishing within them.

Kuaycharoen (2002) examined adaptation of the Nong Nam Yai community property right regime under the influence of both external economic and political changes and internal cultural factors. She demonstrated that the change of the barrage fishery from private to community property had led to a structural change in rights and duties in the relationship between people and resources. The Nong Nam Yai community combines such formal institutions as the Village Committee and Sub-district Administrative Organization with local institutions like a belief in ancestral spirits used in fish conservation zones. Both individual and community rights are recognized. For example, certain parts of ponds are reserved for villagers or groups of villagers; and the right of individuals the locations for installing large fishing gears is recognized via the auction system. Khumsri (2008) and Khumsri et al. (2009) further demonstrated that within a single small geographical area used as fishing ground, fisheries resources in the Lower Songkhram River Basin are managed under a complex and multiple set of overlapping, complementary and conflicting individual, common and state property rights.

1.4.5 Vietnam

Until a decade ago there had been no focused study of pre-existing fisheries management systems (*van chai*) in Vietnam (Ruddle 1994c, 1998a). The study by Ruddle (1998a) was based on extended field research conducted in 1995 and 1996 at eight van chai in five provinces along the coast of the Central and Southern regions. Several articles by Nguyen Duy Thieu (1993, 2002a, b, 2003, 2007) mention pre-existing management based on the van chai system.

A more comprehensive overview of the van chai system is provided by Ruddle and Tuong (2009) which added to the literature those parts of Vietnam where recent research either has been conducted or is on going. In that volume Ha and Nguyen Duy Thieu (2009) describe the historical roots of the institution in rivers in northern Vietnam and coastal lagoons in the Central Region; Nguyen Duy Thieu (2009) further examined floating fishing villages in the social life of fishers in South Central Region; Ruddle and Luong (2009) summarized the history of the van chai in Binh Thuan Province, and analyzed their management and spiritual role, based mainly on the case of Van Thuy Tu, Phan Thiet City; Nguyen Quang Vinh Binh (2009) provided an overview of van chai in Thue Thien Hue; and Tran (2009) examined the role of whales in the spiritual life of coastal fishing villages in Binh Dinh Province.

1.5 Management Functions and Approaches

In any fishery four existing or potential problems require management. These are: (1) resource flows (or the regular availability of harvestable fish); (2) stock externalities (or economic and social impacts of harvesting interactions); (3) technological (gear) externalities (or the incompatibility of various gears); and (4) allocation problems (or competition for access to unevenly distributed resources) (Ruddle 1989b, 1996a, 1998b).

There are two basic ways of addressing these issues. One is the Western scientific fisheries management that focuses on fish stocks and stock externalities, and assumes an open access resource regime. In other words, it focuses on trying to manage what is unknown (and perhaps inherently unknowable) and thus unmanageable.

In contrast, local pre-existing management systems generally take a different approach. They are focused on (1) gear externalities and allocation problems, (2) implemented based on defined the geographical areas and controlled access, (3) self-monitored by local fishers, and (4) enforced by local moral and political authority (Ruddle 1996a).

So, in striking contrast to conventional fisheries management, such local systems focus on human problems that are inherently manageable. This implicitly accounts for the complex multi-species and multi-gear nature of the tropical fisheries resource, thereby avoiding issues lacking inherent solutions (Ruddle 1996a).

It is sobering to realize that such ideas as 'Adaptive Management' and 'The Ecosystem Approach' are commonly portrayed as being of both recent and Western origin, which were long ago incorporated in resource management systems of the Asia-Pacific Region (Ruddle and Hickey 2008). Adaptive Management is portrayed as 'learning by doing'. In other words it is an old trial-and-error management research approach that has been used in the Asia-Pacific Region and elsewhere for centuries, perhaps millennia. The idea of Adaptive Management as re-discovered in Western scientific circles from the late-1970s (Holling 1978; Walters 1986, 1997; Lee 1993), explicitly acknowledges uncertainties and knowledge gaps about the response of a system to management actions. This is important because of the inherent uncertainty or risk involved in any attempt to manage natural resources and the environment, and because scientific knowledge of ecosystem functions is usually not enough to provide unequivocal answers to management problems. Uncertainty is usually ignored and management proceeds on the assumption that its impact on ecosystems and resources is certain. Because this is often not the case, management policy can itself cause environmental degradation, the loss of ecological goods and services, and economic and social instability (Walters 1986; Walters and Holling 1990).

Similarly, 'The Ecosystem Approach' is well known from various places in the Pacific Islands, where it was expressed traditionally in the concept of 'corporate estate' (Ruddle 1994c). In Southeast Asia they occur widely in Indonesia, as in Lombok (Satria 2007 and Chapter 2, this volume) and Maluku (Adhuri 2002 and Chapter 2, this volume). There is nothing new or startling about the Ecosystem Approach, which is simply an outline of an interdisciplinary methodology for environmental research, planning and management, as adopted in 2000 by the Convention on Biological Diversity (CBD 2000). In fact, its basic ideas are inherent in most pre-existing or traditional systems of management that acknowledge ecological relationships. As exemplified from Lombok Island (Satria 2007 and Chapter 2, this volume), the practitioners of pre-existing integrated systems clearly understood that resource enhancement and habitat protection are two interrelated management functions, since stock enhancement is pointless if the habitat(s) on which it depends cannot simultaneously be protected. This brings fishing communities into a relationship with upstream and in situ users of other resources, with the impacts that those resource uses have on the aquatic environments on which stocks depend (Ruddle 2011).

Rural economies in the tropics have never been just about farming. It is noteworthy that throughout most of the Asia-Pacific tropics a non-specialized approach that included farming, fishing and exchange systems was adopted to spread the risk resulting from various threats. The production activities of most tropical rural households consist of several complementary economic activities that as a whole provide a balance of subsistence goods and a spreading of risk. These may be closely linked, as in pre-existing integrated farming systems that combine cropping with animal husbandry and aquaculture or rice field fisheries (Ruddle 1991; Ruddle and Zhong 1988). This means that pre-existing Adaptive Management, as expressed in the 'Estate Concept' and other social arrangements, can be interpreted as a logical attempt to spread risk in an uncertain environment with limited resources. This idea has now been re-packaged by Western academics and donors as the 'Livelihoods Approach'.

In addition, a fragmented view of traditional household economic activities is grossly misleading because individual rural households rarely function without reference to others in the community. Typically, a high degree of interaction exists, and household economics is dominated by tradition, kinship and the community wide needs for security and survival. In the long run, household welfare depends on that of other households, and on such relationships as mutual assistance, welfare and patronage.

1.6 Basic Design Principles of Pre-existing Systems

Both the problems of gear externalities and assignment are overcome in pre-existing fisheries management systems at two levels. This is achieved at the first level by (1) control of a fishing area as a strictly bounded property, and (2) establishing precise social boundaries, by rights, to define who has access rights to that area (Ruddle 1996a, b). At the second level boundaries are set by rules of operational behavior that then specify assignments of time and place within the group having access (Ruddle 1996a, b). The first level is sustained by rights of exclusion, or limited access that maintain the private area of a local community of fishers against outsiders. The second level, intra-group operational rules, is sustained by local authority that has the power to punish offenders (Fig. 1.3).

Several generic key elements or 'Design Principles' characterize any successful local management systems. These are (1) authority or leadership, (2) rights, (3) rules, (4) monitoring, accountability and enforcement, and (5) sanctions (Fig. 1.4).

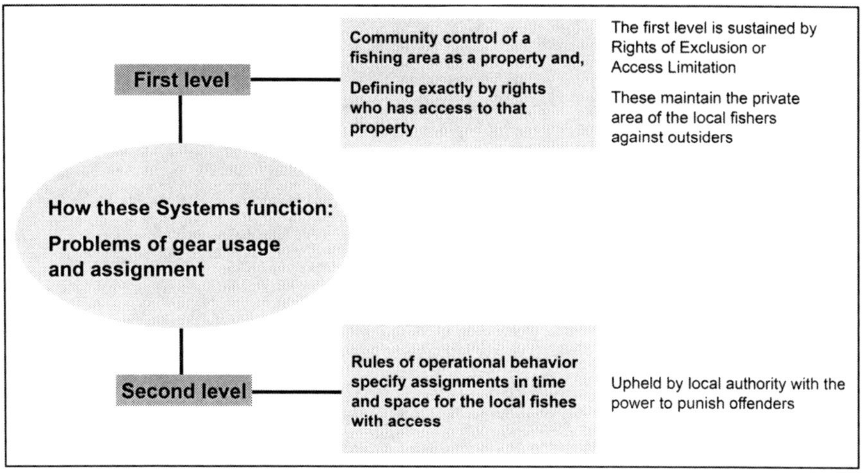

Fig. 1.3 How pre-existing systems function

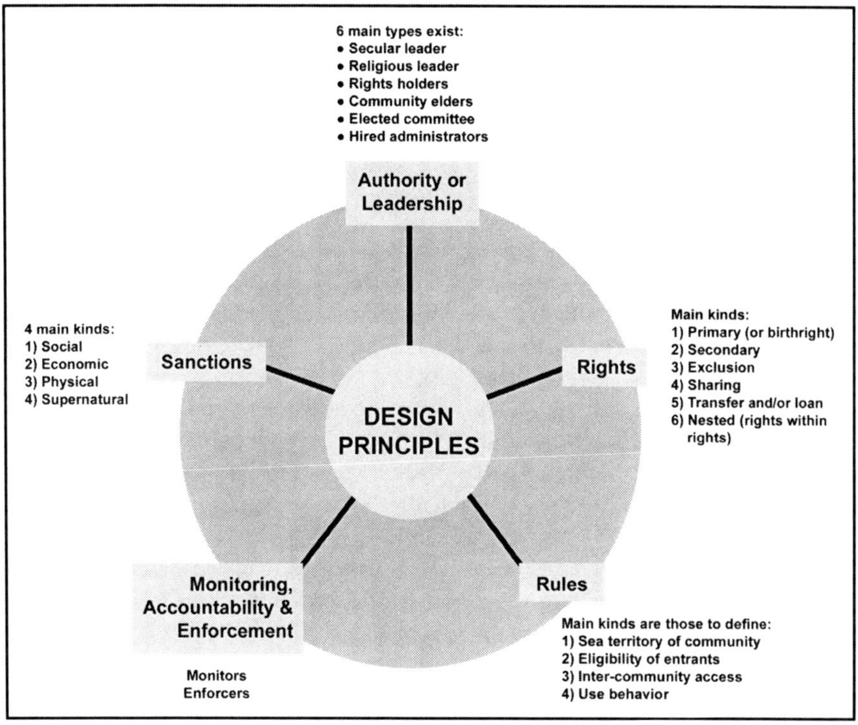

Fig. 1.4 The design principles of pre-existing systems

1.6.1 Authority or Leadership

In pre-existing community based resource management systems, control and management is usually vested in traditional authority, the nature of which varies according to social organization. Four principal types can be recognized in the Asia-Pacific region: secular leaders, religious leaders, specialists, and rights owners. These categories frequently overlap, and responsibility is divided and shared.

In many societies a group of traditional secular leaders or an organization, usually some kind of 'village council', manages marine resources by regulating the use of community sea space and protecting resources against over-exploitation. Often, however, land and sea is disposed of by a chief, who exercises his authority on behalf of the entire community. The role of religious leaders in pre-existing systems of resource management is also widespread in Asia. These can be both traditional religious leaders, as in Indonesia (Adhuri 2002; Satria 2007), or church officials, as with the Roman Catholic Church in Sri Lanka (Atapattu 1987). Commonly, too, marine resources are managed by fisheries specialists, or 'master fishermen', who function under some form of higher authority. Finally, rights-holders themselves commonly have management authority over marine resources. Frequently, this level of authority is vested in the senior person of a lineage, family, or other small social group, as in the Kei Islands of Indonesia (Adhuri 2002; Chapter 2 this volume).

1.6.2 Rights

Under pre-existing management systems, the exploitation of aquatic resources is governed by use rights to a property, the claim to the resources or services of which is protected by both local or customary law and practice. Such rights define the uses legitimately viewed as exclusive, as well as the penalties for violating the rights. Common characteristics are exclusivity, the right to determine who can use a fishing ground, transferability, the right to sell, lease, or bequeath the rights, enforcement, and the right to apprehend and penalize violators. The right of enforcement that enables exclusion of the free-riding outsider is a key characteristic, because without it all other rights are diminished either actually or potentially.

Rights to aquatic resources in pre-existing systems may be primary or secondary. They may be further classified into rights of occupation and use. The relationship between the two main types, primary and secondary, is an important and complex characteristic of many pre-existing management systems, in which overlapping and detailed regulations on the use of technologies and particular species are widespread. Individual rights as sub-divisions 'nested within' holdings occur, as do rights of transfer, loan and sharing. Almost universally members of fishing communities have primary resource rights by virtue of having been born into that social group. Such rights to exploit fisheries are subject to various degrees of exclusiveness, which depends on community social organization and local culture. Most commonly, pre-existing fisheries rights apply to areas. However, superimposed on these may be claims held by individuals or groups to a particular species or to a specific fishing technology.

Primary rights are usually those to which a group or an individual is entitled via inheritance (i.e., a birthright), by direct links to the core of a descent based corporate group. They are generally comprehensive, since only they confer access to all resources within a defined territory. Inheritance, ancestral interests, social obligations, and cooperative relationships within a social group provide continuity of ownership and rights.

Secondary rights are more limited than primary rights, often being restricted to specific fishing methods. They are acquired through affiliation with a corporate group, by marriage, traditional purchase, exchange, as a gift, or as reciprocity for services. Sometimes they may be inherited.

In some societies rights to fisheries, which are usually to areas, are overlain by other rights, generally those to species and those to gear types. Many such nested rights are quite simple, like those to locations for large fixed gear. However, complex cases occur, as in the Lower Songkhram River Basin of Northeastern Thailand, where concurrent management under Department of Fishery Regulations and by local communities based on pre-existing rules that recognize individuals have ownership of fishing and exclusion rights in such areas results in a complex set of overlapping, complementary and conflicting individual, common and state property rights within a single, small fishing ground (Khumsri et al. 2009).

Some pre-existing management systems permit the permanent, temporary, or occasional transfer of rights to other social units. Often, temporary and occasional transfer requires users to compensate rights-owners in cash or, more commonly, in kind, usually with a portion of the catch. In other societies, like Japan (Ruddle 1987), however, individual fishermen are proscribed by either statutory or customary law from transferring their rights.

In some places areal rights are shared between or among different corporate communities. Commonly shared rights have deep historical roots, and invariably sharing is done only for the most productive waters or where kinship ties are strong.

1.6.3 Rules

Rules define how a property right is to be exercised, by specifying required, permitted and forbidden acts in exercising the authority provided by the right. Whereas a right authorizes a fishermen to work a specific fishing ground, his options in exercising it are governed by rules which may, for example, specify gear type used or seasonal restrictions, among other limitations. The more complete a set of rights, the less exposed are fishermen to the actions of others.

Basic rules define the geographical areas to which rights are applied, define those persons eligible to fish within a community's aquatic area, and govern access of outsiders. Operational rules govern fishing behavior, gear externalities, assignment issues (temporal allocation, species rules, as well as specify unacceptable fishing behavior, conservation practices, and distribution of the catch within the community. Rules defining access to harvested fish are widespread. These are an extremely important set of rules in many societies since, in terms of equity within a community, access to fish once harvested can be as or more important than access to fishing grounds (Collier et al. 1979; Kendrick 1993). Such rules include those to provision the family and community, those required as subsequent and continual repayment for the acquisition of fishing rights, and those enmeshed in general community sharing and reciprocity and related norms concerning equity and fairness (Ruddle 1994c).

1.6.4 Monitoring, Accountability and Enforcement

If rights are to be meaningful, provision must be made within the system for monitoring compliance with rules and imposing sanctions on violators. Under pre-existing aquatic resource management systems monitoring and enforcement are generally undertaken within the local community; resource users policing themselves, and being observed by all others as they do so.

1.6.5 Sanctions

Sanctions are widely invoked for the infringement of fisheries rights and the breaking or ignoring of locally formulated rules governing fishing and other uses of aquatic resources. Four principal types of sanctions are widely invoked; social, economic, physical punishment, and supernatural. Social sanctions include ridicule, shaming, ostracism, and banishment. Economic sanctions include monetary and in-kind fines, destruction of gear and forced labor, among others. Physical punishment was widely applied for the violation of rules. Supernatural sanctions are not uncommon, and fear of them reinforces the other types of sanction.

In summary, it is fundamental to appreciate that pre-existing systems of fisheries management are based on a specific aquatic area that comprises a common property of the fishing community. Common properties have peculiar characteristics that entail rights and duties (a reciprocal of a right). The main ones are that the common property is (1) co-owned by individual fishers in their position as members of a recognized group; (2) that these co-owners comprise the management groups; (3) that these co-owners have the right to exclude non-members; and (4) that these co-owners have both rights and duties regarding use rates and maintenance of the resource owned.

Such pre-existing systems of property rights and associated regimes of rights and rules closely reflect social organization and local power structure, and seem not to have been based principally on ecological conditions, which would be the case were their primary purpose resource conservation. Rather, as would be expected, since property is a social relationship that defines its holder's security of claim to a resource or to the services or benefits it provides, they reflect a correlation among property, property rights, and social organization (Ruddle 1989a). Management systems in the aquatic domain often, but not always, mirror those on land.

1.7 Success Stories

As demonstrated by Johannes' (1977, 1978) pioneering work in the Pacific Islands, 40 years ago the centuries-old practice of marine resource management was declining in the face of Westernization. However, from the 1980s this trend had reversed in several countries. This resurgence of pre-existing practices continues apace, based on limited entry, protected areas, closed areas, closed seasons, and strict control of either damaging or particularly efficient gear, among other methods (Johannes 2002).

A particularly striking success was recorded in Vanuatu during the period 1990–1993, when a massive resurgence of village-based marine resource management occurred, based on pre-existing methods. This was orchestrated by the Fisheries Department. Only one village of the 26 surveyed by Johannes (1998) had not introduced the new measures between 1990 and late-1993. Although government assistance and advice concerned only *Trochus* sp., the success of conservation measures designed specifically for it encouraged villagers to introduce controls for many other species of fish and invertebrates. The 26 villages introduced some form of explicitly conservation-based taboos on their fishing grounds, and all asserted their right to exclude outsiders from them. Johannes' interviews revealed that, although traditional in times past, these fishing taboos were widely reintroduced and had been applied for the first time in living memory within a space of only three-and-a-half years.

A similar resurgence of also occurred in Samoa during the 1990s. There again it was impelled by government, when the Fisheries Division assisted villagers to design and implement a legal device to overcome their inability to stop poaching on village fishing grounds. In pre-colonial times there existed in Samoa a strong system of village marine tenure. But his broke down as the colonial rulers transferred ownership of marine waters to the State. As a result, although village chiefs could control their own villagers, they had no power over the actions of outsiders. Reciprocal access rights had once existed, when marine stocks were abundant, but population growth and resource depletion put an end to that. In the late-1980s the problem was addressed by passage of the Fisheries Act (1988), which established a process whereby any village regulation could become legally recognized. Traditional authority was also reinforced by the Village Fono (Council of Chiefs) Bill (1990), which amended the constitution to give chiefly authority in accordance with Samoan custom over nearshore fisheries, to which village rights were given primacy. Thus the incentive for villagers to manage their fishing grounds was restored.

The situation is not so advanced in Southeast Asia. In Indonesia, as demonstrated in Chapter 2 of this volume, after having been delegitimized for three decades under the 'New Order Regime' (1966–1998), pre-existing systems were revived by local initiative. This is exemplified by the *awig–awig* of Lombok Island. Nevertheless, legislation in Indonesia has not addressed explicitly the issue of pre-existing fisheries management systems. In both Laos (Chapter 3, this volume) and Thailand (Chapter 5, this volume) legislation has not addressed directly pre-existing systems, although their local importance is now evident. In the Philippines (Chapter 5, this volume) whereas claims like that of the Tagbanua of Palawan Island can be made according to the Indigenous Peoples' Rights Act of 1977, government does little or nothing to uphold them against various powerful interests that seek to violate or overturn them. In Vietnam (Chapter 6, this volume) a modern management role has been advocated for the pre-existing van chai system (Ruddle and Tuong 2009), and is now being examined.

1.8 Contents of This Book

Regarding pre-existing systems of marine resources management, Ruddle (1994c) observed that in both continental and insular Southeast Asia the tenurial relationship of small-scale fishermen to resource areas and resources is not well known, and that only vestiges of what probably were more widespread systems remained. Ruddle (1994c) noted that both historical and contemporary reports of systems from continental Southeast and East Asia were conspicuously absent, and that field research and archival study of the records of former colonial administrations is likely to prove profitable in filling these gaps. However, it should be understood that reference was being made only to marine systems, whereas in continental Southeast Asia inland fisheries have always been more important that their marine counterparts, which historically are a relatively recent development, as in Thailand and Vietnam (Ha and Nguyen 2009).

As demonstrated in this book by the chapters on Laos, Thailand and Vietnam, the situation was different in inland waters, and pre-existing systems were old established in them. Thus in Chapter 3, using case studies of three pre-existing tenure systems for fisheries management from Champasak Province, Ian Baird debunks the myth that all fisheries resources in southern Laos were historically 'open access'. For all three of these ecologically and socially very different examples private resource ownership is socially and culturally sanctioned as part of a common property management system, based on first claims to fishing sites and labor inputs. 'Open access' rarely exists in southern Laos, which may not become evident without sustained observations of different fisheries.

Recent field research, in the Lower Songkhram River Basin (LSRB) of Northeastern Thailand (Khumsri 2008; Khumsri et al. 2009) revealed fisheries resources are managed concurrently by local communities, based on pre-existing or de facto rights, and de jure by the Department of Fishery (DOF), according to the Fisheries Law of 1947. Further, according to the Thai Civil and Commercial Law of 1925, natural resources used in common, such as shores, streams and lakes, are state property (RTG 1930). However, concurrently local communities recognize that individuals have ownership of fishing rights in such areas, and that they also have the right to exclude others from fishing within them. The result is a complex and multiple set of overlapping, complementary and conflicting individual, common and state property rights within a single, small geographical area used as a fishing ground. However, as Khumsri demonstrates in Chapter 5, the performance and sustainability of the present joint system of management is constrained both by a lack of clearly defined property rights and rules aimed specifically at sustainable resource use, and a mismatch between local and state institutional arrangements for fisheries management.

As examined in Chapter 6, in Vietnam the pre-existing fisheries management system known as the van chai emerged from the administrative structure of farming villages in the northern provinces of the country, so their administration and social management reflected traditional Vietnamese agrarian culture. Originally, the van chai administered inland fisheries. However, with the gradually settlement of ethnic Vietnamese along the coast of the Central Region, where inland fisheries were insignificant but marine fisheries became of major importance, the van chai was adapted to the needs of marine fishing communities, where it became the focus for spiritual activities related to fishing. Each new fishing community along the South-Central Coast established a van chai to worship the Whale God.

The recent disappearance of pre-existing systems is one reason why they have not been documented. In the case of Indonesia, pre-existing systems of fisheries management were delegitimized during the 'New Order Era' (1966–1998), before their importance was recognized. They were revived after a 30-year hiatus, when the 'Reform Era' began, in 1998. One such system is the *awig–awig* and *sawen* of North Lombok, and another is the petuanan and sasi of Maluku. These are examined in Chapter 2. Local people revived three awig–awig and adapted them to the contemporary need of overcoming destructive fishing practices and implementing a system for sustainable fisheries management.

Another common reason for the failure to document pre-existing systems is the absence of recent field research, particularly in remote areas, combined with the common assumption that such systems either do not exist in a given area or that they have little or no relevance to modern fisheries management. Such is the case examined in Chapter 4 of the *mataw* fishers of Batanes, the ten small and northernmost islands of the Philippine archipelago, who engage in the seasonal capture of Flying fish and Dorado. Each fisher is identified by the '*vanua*' or 'port', a specific spatial location to which he belongs.

Finally, as a group these cases highlight some important and generally overlooked aspects of the characteristics and context of pre-existing systems that usually escape attention. These are their fundamental role in the management of fishing communities; the existence of multiple, overlapping, flexible and adaptable rights;

that they are often involved a set of human ecosystems and their resources, and not just fisheries, which are managed in a coordinated manner; and that pre-existing systems are greatly affected by a constellation of interacting external pressures for change. In overlooking these, policy makers and planners commit a serious error of judgment.

References

Adhuri, D. S. (2002). Selling the sea, fishing for power: a study of conflict over marine tenure in the Kei Islands, Eastern Indonesia. Dissertation, The Australian National University.

Anon. (1991). *Laporan penelitian hak adat kelautan di Maluku*. Ambon: Yayasan Hualopo, Fak. Hukum dan Pusat Studi Maluku Unpatti.

Aswani, S., & Hamilton, R. (2004). Integrating indigenous ecological knowledge and customary sea tenure for conservation of Bumphead Parrotfish (*Bolbometopon muricatum*) in the Roviana Lagoon, Solomon Islands. *Environmental Conservation, 31*(1), 1–15.

Atapattu, A. R. (1987). Territorial use rights in fisheries (TURFs) in Sri Lanka: Case studies on Jakottu fisheries in the Madu Ganga estuary and Kattudel fishery in the Negombo Lagoon. In Anon (Ed.), *Papers presented at the Symposium on the exploitation and management of marine fishery resources in Southeast Asia held in conjunction with the 22nd session of the IPFC Darwin, 16–26 February, 1987.RAPA Report:1987/10.* (pp. 379–401). Bangkok: Regional Office for Asia and the Pacific (RAPA), FAO.

Bailey, C., & Zerner, C. (1992). Local management of fisheries resources in Indonesia: Opportunities and constraints. In R. B. Pollnac, C. Bailey, & A. Poernomo (Eds.), *Contributions to fishery development policy in Indonesia* (pp. 38–56). Jakarta: Central Research Institute for Fisheries, Agency for Agricultural Research and Development, Ministry of Agriculture.

Baird, I. G. (1999, August). Towards sustainable co-management of Mekong River inland aquatic resources, including fisheries, in southern Lao PDR. Paper presented at the International Workshop on Fisheries Co-management. Penang, Malaysia.

Baird, I. G. (2001). Towards sustainable co-management of Mekong River aquatic resources: The experience in Siphandone wetlands. In G. Daconto (Ed.), *Siphandone wetlands* (pp. 89–111). Bergamo, Italy: CESVI.

Baird, I. G. (2006a). Strength in diversity: Fish sanctuaries and deep-water pools in Lao PDR. *Fisheries Management and Ecology, 13*, 1–8.

Baird, I. G. (2006b). *Probarbus jullieni* and *Probarbus labeamajor*: The management and conservation of two of the largest fish species in the Mekong River in southern Laos. *Aquatic Conservation: Marine and Freshwater Ecosystems, 16*, 517–532.

Baird, I. G., Flaherty, M. S., & Phylavanh, B. (2003). Rhythms of the river: Lunar phases and migrations of small carps (Cyprinidae) in the Mekong River. *Natural History Bulletin of the Siam Society, 51*(1), 5–36.

Benda-Beckmann, F. von, Benda-Beckmann, K. von, & Brouwer, A. (1992, August). *Changing 'Indigenous Environmental Law' in the Central Moluccas: Communal regulation and privatization of sasi.* Paper presented at the Congress of the Commission on Folk Law and Legal Pluralism. Wellington, New Zealand.

Berkes, F. (1999). *Sacred ecology: Traditional ecological knowledge and resource management.* Philadelphia: Taylor and Francis.

Blair, E. H., & Robertson, J. A. (1903–1909). *The Philippine Islands: 1493–1898.* Cleveland, OH: A.H.Clark Co.

Bromley, D. W. (1992). The commons, property, and common-property regimes. In D. W. Bromley (Ed.), *Making the commons work: Theory, practice and policy* (pp. 3–15). San Francisco, CA: Institute for Contemporary Studies.

Cantero-Pastrano, C. L. (1955). Report on a Visayan fishing barrio. *Philippine Sociological Review, 3*, 15–22.

CBD (Convention on Biological Diversity). (2000). *Decisions adopted by the Conference of the Parties to the Convention on Biological Diversity at its fifth meeting, Nairobi, 15–26 May 2000 Decision V/6, UNEP/CBD/COP/5/23*. Nairobi: UNEP.

Chong, K.-C., Smith, I. R., & Lizarondo, M. S. (1982). Economics of the Philippine milkfish resource system. *Resource Systems Theory and Methodology Series 4*. Tokyo: The United Nations University.

Claridge, G., Sorangkhoun, T., & Baird, I. G. (1997). *Community fisheries in Lao PDR: A survey of techniques and issues*. Vientiane: The International Union for Conservation of Nature and Natural Resources (IUCN).

Collier, W. L., H. Hadikoesworo, H. & Malingreau, M. (1979). Economic development and shared poverty among Javanese sea fishermen, In Librero, A.R. & Collier, W.L. (Eds.), *Economics of aquaculture, sea-fishing and coastal resource use in Asia. Proceedings of the 2nd Bienniel Meeting of the Agricultural Economics Society of Southeast Asia, Nov. 3–6, 1977, Tigbauan, Iloilo, Philippines* (pp. 218–236). Bangkok/Los Baños: Agricultural Development Council and Philippine Council for Agriculture and Resources Research.

Cordell, J. C. (1977). Carrying capacity analysis of fixed territorial fishing. *Ethnology, 17*(1), 1–24.

Cruz, W. D. (1982). Institutional and technical aspects of access to municipal fishery resources. In I. R. Smith, & A. N. Mines (Eds.), *Small-scale fisheries of San Miguel Bay, Philippines: Economics of production and marketing*. ICLARM Technical Reports 8 (pp. 95–103). Quezon City, Manila/Tokyo: Institute of Fisheries Development, College of Fisheries, University of the Philippines in the Visayas, ICLARM, and the United Nations University.

Cruz, W. D. (1986). Overfishing and conflict in a traditional fishery: A resource management dilemma. In Anon (Ed.), *Proceedings of the conference on common property resource management, 21-26/4/1985* (pp. 115–135). Washington D.C.: National Academy Press.

Davis, A., & Ruddle, K. (2010). Creating confidence: Rational skepticism and systematic enquiry in local ecological knowledge research. *Ecological Applications, 20*(3), 880–894.

de Castro, F., & Begossi, A. (1995). Ecology of fishing on the Grande River (Brazil): Technology and territorial rights. *Fisheries Research, 23*, 361–373.

Durning, A. (1992). *Guardians of the land: Indigenous peoples and the health of the earth, Worldwatch Paper 112*. Washington, DC: Worldwatch Institute.

Fa'asili, U., & Kelokolo, I. (1999). The use of village bylaws in marine conservation and fisheries management. *SPC Traditional Marine Resource Management and Knowledge Information Bulletin, 11*, 7–10.

Galvez, R. E. (1991). Some socioeconomic issues in artificial reefs management: A case of Lingayen Gulf, Philippines. *Tropical Coastal Area Management, 6*(1–2), 6–7.

Ha, X. T., & Nguyen, D. T. (2009). The *van chai* and its role in the hierarchy of fisheries administration in Vietnam. In K. Ruddle & P. L. Tuong (Eds.), *The van chai of Vietnam: Managing nearshore fisheries and fishing communities* (pp. 21–35). Hong Kong: International Resources Management Institute.

Hardin, G. (1968). The tragedy of the commons. *Science, 162*, 1234–1248.

Hart, D. V. (1956). *Securing aquatic products in Siaton Municipality, Negros Oriental Province, Philippines. Monograph 4*. Manila: Institute of Science and Technology.

Hickey, F. R. (2006). Traditional marine resource management in Vanuatu: Acknowledging, supporting and strengthening indigenous management systems. *SPC Traditional Marine Resource Management and Knowledge Information Bulletin, 20*, 11–23.

Holling, C. S. (Ed.). (1978). *Adaptive environmental assessment and management*. New York: Wiley.

Johannes, R. E. (1977). Traditional law of the sea in Micronesia. *Micronesica, 13*, 121–127.

Johannes, R. E. (1978). Traditional marine conservation methods in Oceania and their demise. *Annual Review of Ecology and Systematics, 9*, 349–64.

Johannes, R. E. (1981). *Words of the Lagoon: Fishing and marine lore in the Palau District of Micronesia*. Berkeley, CA: University of California Press.

Johannes, R. E. (1994). Pacific Island peoples' science and marine resource management. In J. Morrison, P. Geraghty, & L. Crowl (Eds.), *Science and the Pacific Island peoples* (Ocean and coastal studies, Vol. 1, pp. 81–90). Suva: Institute of Pacific Studies.

Johannes, R. E. (1998). Government-supported, village-based management of marine resources in Vanuatu. *Ocean and Coastal Management, 40*, 165–186.

Johannes, R. E. (2002). The renaissance of community-based marine resource management in Oceania. *Annual Reviews of Ecology and Systematics, 33*, 317–40.

Johannes, R. E. (2003). Use and misuse of traditional ecological knowledge and management practices. In D. G. Dallmeyer (Ed.), *Values at Sea: Ethics for the marine environment* (pp. 111–126). Athens GA: University of Georgia Press.

Johannes, R. E., & Hickey, F. R. (2004). *Evolution of village-based marine resource management in Vanuatu between 1993 and 2001. Coastal Region and Small Island Papers 15*. Paris: UNESCO.

Kada, Y. (1984). The evolution of joint fisheries rights and village community structure on Lake Biwa, Japan. In K. Ruddle & T. Akimichi (Eds.), *Maritime institutions in the Western Pacific. Senri ethnological studies, 17* (pp. 137–158). Osaka: National Museum of Ethnology.

Kalagayan, B. N. V. (1991). Philippine initiatives in marine coastal conservation. *Lundayan, 1st quarter*, 2–7.

Kendrick, A. (1993, June). *The erosion and relocation of local resource management institutions in a Javanese fishery.* Paper presented at the 4th Annual Meeting of the International Association for the Study of Common Property, Manila.

Khumsri, M. (2008). Co-Management institutions for sustainable inland fisheries management: A case study of the Lower Songkhram River Basin, Thailand. Dissertation, Asian Institute of Technology.

Khumsri, M., Ruddle, K., & Shivakoti, G. P. (2009). Rights and conflicts in the management of fisheries in the Lower Songkhram River Basin, Northeast Thailand. *Environmental Management, 43*, 557–570.

Kissya, E. (1995). *Sasi aman haru-ukui: Traditional management of sustainable natural resources in Haruku*. Jakarta: Sejati Foundation.

Kuaycharoen, P. (2002). *Common property rights regimes: dynamics of management of freshwater fisheries resources in the communities of the Lower Songkhram River Basin.* Masters Research Study No 974-327-827-3. Thesis, Thammasat University (in Thai).

Lee, K. N. (1993). *Compass and gyroscope: Integrating science and politics for the environment.* Washington, DC: Island Press.

Levieil, D. P. (1987). Territorial use-rights in fishing (TURFS) and the management of small-scale fisheries: The case of Lake Titicaca. Dissertation, University of British Columbia.

Lokollo. (1988). *Hukum sasi di Maluku: Suatu potret binamulia lingkungan pedesaan yang dicari pemerintah.* Ambon: Faculty of Law, University of Pattimura (in Indonesian).

Lopez, M. D. G. (1985). Notes on traditional fisheries in the Philippines. In K. Ruddle & R. E. Johannes (Eds.), *The traditional knowledge and management of coastal systems in Asia and the Pacific* (pp. 190–206). Jakarta: UNESCO-ROSTSEA.

McGrath, D. G., de Castro, F., Futemma, C., de Amaral, B. D., & Calabria, J. (1993). Fisheries and the evolution of resource management on the lower Amazon floodplain. *Human Ecology, 21*(2), 167–195.

Naamin & Badrudin (1992). The role of coastal village communities and fishermen's organization in the management of coastal fisheries resources in Indonesia. In Anpon. (Ed.), *Proceedings of FAO/Japan Expert Consultation on the Development of Community-Based Coastal Fishery Management System for Asia and the Pacific Volume 2*. Rome: FAO.

Neiland, A. E., Weeks, J., Madakan, S., & Ladu, B. (1994). *Traditional fisheries jurisdiction in N E Nigeria: Results of a survey in 1994 at Lake Chad, River Benue and the Nguru-Gashua wetlands, Research Paper 72*. Portsmouth, UK: Centre for the Economics and Management of Aquatic Resources, University of Portsmouth.

Nguyen, D. T. (1993). A few features on the way of life of the Bo Lo and other fishing inhabitants at Cua Sot estuary. *Vietnam Social Sciences, 4*, 158–211.

Nguyen, D. T. (2002a). *Fishing communities in Vietnam*. Hanoi: Social Science Publishing House (in Vietnamese).

Nguyen, D. T. (2002b). Religions and beliefs in Vietnamese fishing communities. *Ethnology Magazine, 1*, 62–71.

Nguyen, D. T. (2003). Sea daily routes of the Bo Lo in the Northern area of the central region. *Magazine Folklore Culture Review, 6*, 27–32 (in Vietnamese).

Nguyen, D. T. (2007). Studies of fishing communities in Vietnam. *Anthropology Review, 2*, 45–57.

Nguyen, D. T. (2009a). The role of floating fishing villages in the social life of fishers in the South-Central Region. In K. Ruddle & P. L. Tuong (Eds.), *International resources management institute* (pp. 36–46). Hong Kong: FAO.

Nguyen, Q. V. B. (2009b). *Van chai* in Thua Thien Hue: Tradition, present and future challenges. In K. Ruddle & P. L. Tuong (Eds.), *The van chai of Vietnam: Managing nearshore fisheries and fishing communities* (pp. 66–76). Hong Kong: International Resources Management Institute.

Ostrom, E. (1990). *Governing the commons: The evolution of institutions for collective action*. New York: Cambridge University Press.

Ostrom, E. (1998). *The comparative study of public economies, Acceptance paper delivered by the 1997 recipient of The Frank E. Seidman distinguished Award in Political Economy*. Memphis, TN: P. K. Seidman Foundation.

Pannell, S. (1997). Managing the discourse of resource management: The case of sasi from 'Southeast' Maluku, Indonesia. *Oceania, 67*, 289–307.

Pauly, D. (1994). May apply in the tropics – but not here!'. In D. Pauly (Ed.), *On the sex of fish and the gender of scientists: Essays in fisheries science* (pp. 3–6). London: Chapman & Hall.

Polunin, N. V. C. (1984). Do traditional marine "reserves" conserve? A view of Indonesian and Papua New Guinean evidence. In K. Ruddle & T. Akimichi (Eds.), *Maritime institutions in the Western Pacific. Senri ethnological studies* (pp. 267–283). Osaka: National Museum of Ethnology.

Roberts, T. R., & Baird, I. G. (1995). Traditional fisheries and fish ecology on the Mekong River at Khone Waterfalls in southern Laos. *Natural History Bulletin of the Siam Society, 43*, 219–262.

RTG (Royal Thai Government). (1930). *Thai civil and commercial law of 1925*. Bangkok: Royal Thai Government (in Thai).

Ruddle, K. (1987). *Administration and Conflict Management in Japanese Coastal Fisheries*, FAO Fisheries Reports 273. Rome: FAO.

Ruddle, K. (1989a). Social principles underlying traditional inshore fishery management systems in the Pacific Basin. *Marine Resource Economics, 5*(4), 231–243.

Ruddle, K. (1989b). Traditional sole property rights and modern inshore fisheries management in the Pacific Basin. In H. Campbell, K. Menz, & G. Waugh (Eds.), *Proceedings of an international conference held at Hobart, Tasmania, 20-22/3/1989* (pp. 68–76). Canberra: Australian Centre for International Agricultural Research.

Ruddle, K. (1991). Integrated farming systems and future directions for Asian farming systems research and extension. *Journal of the Asian Farming Systems Association, 1*(1), 91–99.

Ruddle, K. (1994a). External forces and change in traditional community-based fishery management systems in the Asia-Pacific Region. *Maritime Anthropological Studies, 6*(1–2), 1–37.

Ruddle, K. (1994b). Marine tenure in the 90s. In G. R. South, D. Goulet, S. Tuqiri, & M. Church (Eds.), *Traditional Marine tenure and sustainable management of marine resources in Asia and the Pacific: Proceedings of the international workshop 4th–8th July, 1994* (pp. 6–45). Suva: The International Ocean Institute, South Pacific and Marine Studies Programme, University of the South Pacific.

Ruddle, K. (1994c). *A Guide to the Literature on Traditional Community-based Fishery Management in the Asia-Pacific Tropics*. Fisheries Circular No. 869, FIPP/C869. Rome: FAO.

Ruddle, K. (1995). The role of validated local knowledge in the restoration of fisheries property rights: the example of the New Zealand Maori. In S. Hanna & M. Munasinghe (Eds.), *Property rights in a social and ecological context: Vol. 2 Case studies and design applications* (pp. 111–120). Stockholm/Washington, DC: The Beijer International Institute of Ecological Economics/The World Bank.

Ruddle, K. (1996a). Traditional management of reef fishing. In N. V. C. Polunin & C. Roberts (Eds.), *Reef fisheries* (pp. 315–335). London: Chapman & Hall.

Ruddle, K. (1996b). Boundary definition as a basic design principle of traditional fishery management system in Pacific Islands. *Geographische Zeitschrift, 84*(2), 94–102.

Ruddle, K. (1998a). Traditional community-based coastal marine fisheries management In Viet Nam. *Ocean and Coastal Management, 40*(3), 1–22.

Ruddle, K. (1998b). The context of policy design for existing community-based fisheries management systems in the Pacific Islands. *Ocean and Coastal Management, 40*(4), 105–126.

Ruddle, K. (2007a). Wronging rights and righting wrongs. In W. Taylor, M. Schechter, & L. Wolfson (Eds.), *Globalization: Effects on fisheries resources* (pp. 456–483). Cambridge: Cambridge University Press.

Ruddle, K. (2007b). Fishing rights: misconceptions, outright prejudice. Samudra The Triannual Journal of the International Collective in Support of Fishworkers, Report No 48: 4–9

Ruddle, K. (2007c). Introduction to the collected works of R.E. Johannes, publications on marine traditional knowledge and management In K. Ruddle (Ed.), *The collected works of R.E. Johannes: Publications on marine traditional knowledge and management* (pp. 1–17). Hong Kong: International Resources Management Institute.

Ruddle, K. (2011). Coastal-marine resource use in human ecological context: the scale and modes of integration, In V. Christensen, & J. Maclean (Eds.), *Ecosystem Approaches to Fisheries: a Global Perspective*. Cambridge: Cambridge University Press. (in press).

Ruddle, K., & Akimichi, T. (Eds.). (1984). *Maritime institutions in the Western Pacific. Senri ethnological studies, 17*. Osaka: National Museum of Ethnology.

Ruddle, K., & Hickey, F. R. (2008). Accounting for the mismanagement of tropical nearshore fisheries. *Environment, Development and Sustainability, 10*(5), 565–589.

Ruddle, K., & Johannes, R. E. (Eds.). (1985). *The traditional knowledge and management of coastal systems in Asia and the Pacific*. Jakarta: UNESCO-ROSTSEA.

Ruddle, K., & Luong, T. S. (2009). *Van chai* in fisheries village development and management in Binh Thuan Province. In K. Ruddle & P. L. Tuong (Eds.), *The van chai of Vietnam: Managing nearshore fisheries and fishing communities* (pp. 47–65). Hong Kong: International Resources Management Institute.

Ruddle, K., & Tuong, P. L. (Eds.). (2009). *The van chai of Vietnam: Managing nearshore fisheries and fishing communities*. Hong Kong: International Resources Management Institute.

Ruddle, K., & Zhong, G. F. (1988). *Integrated agriculture-aquaculture in South China: The dike-pond System of the Zhujiang Delta*. Cambridge: Cambridge University Press.

Sand, P.H. (1970), Survey of the water resources of the Chad Basin for development purposes. Eleventh Meeting of the Chad Basin Commission, Niamey, Niger, 22–27 June 1970. FAO: Rome.

Santos, A. B. (1980). *Relevant laws on municipal fisheries*. Manila: Bureau of Fisheries and Aquatic Resources.

Sarch, M. R. (1994). *Traditional management of artisanal fisheries, Northern Nigeria, working paper investigating fisheries management systems*. Portsmouth, UK: Centre for the Economics and Management of Aquatic Resources, University of Portsmouth.

Satria, A. (2007). Sawen: institution, local knowledge and myths in fisheries management in North Lombok, Indonesia. In N. Haggan, B. Neis, & I. G. Baird (Eds.), *Fishers' knowledge in fisheries science and fisheries management* (pp. 199–220). Paris: UNESCO.

Satria, A., Umbari, A., Fauzi, A., Purbayanto, A., Sutarto, E., Muchsin, I., et al. (2002). *Menuju Desentralisasi Kelautan*. Jakarta: Cidesindo.

Seibel, H. D. (1994). From cheap credit to easy money: How to undermine rural finance and development. In F. J. A. Bouman & O. Hospes (Eds.), *Financial landscapes reconstructed: the fine art of mapping development* (pp. 19–33). Boulder: Westview Press.

Shoemaker, B., Baird, I. G., & Baird, M. (2001). *The people and their river: A survey of river-based livelihoods in the Xe Bang Fai River Basin in Central Lao PDR*. Vientiane: Lao PDR/ Canada Fund for Local Initiatives.
Silvano, R. A. M., & Begossi, A. (1998) The artisanal fishery of Piracicaba River (São Paulo, Brazil): Fish landing composition and environmental alterations. *Proceedings of the Nineth International Congress of European Ichthyologists* (Ce19), Trieste, Italy. Italian Journal of Zoology, **65**, 527–531.
Silvano, R. A. M., & Begossi, A. (2001). Seasonal dynamics of fishery at the Piracicaba River (Brazil). *Fisheries Research, 51*(3), 59–86.
Smith, I. R. (1981). *The conomics of the milkfish fry and fingerling industry of the Philippines. Technical Report 1*. Manila: ICLARM.
Smith, I. R., & Panayotou, T. (1987). Territorial Use Rights and economic efficiency. In Anon. (Ed.), *Papers presented at the Symposium on the Exploitation and Management of Marine Fishery Resources in Southeast Asia, Darwin, 16–19 February, 1987. RAPA Report: 1987/10*, Indo-Pacific Fishery Commission (pp. 350–366). Bangkok: FAO.
Spoehr, A. (1980). *Protein from the sea: Technological change in Philippine capture fisheries. Ethnology monograph 3*. Pittsburgh: University of Pittsburg.
Tiraa, A. (2006). Ra'ui in the Cook Islands – today's context in Rarotonga. *SPC Traditional Marine Resource Management and Knowledge Information Bulletin, 19*, 11–15.
Tran, V. V. (2009). Whales in the spiritual life of coastal fishing villages in Binh Dinh Province. In K. Ruddle & P. L. Tuong (Eds.), *The van chai of Vietnam: Managing nearshore fisheries and fishing communities* (pp. 77–85). Hong Kong: International Resources Management Institute.
Tubtim, N. (2006). Exclusion, accommodation and community-based natural resource management: legitimizing the enclosure of a community fishery in southern Laos. In S. Tyler (Ed.), *Communities, livelihoods, and natural resources: Action research and policy change in Asia* (pp. 1–17). Ottawa: IDRC 2006.
Tubtim, N., & Hirsch, P. (2005). Common property as enclosure: a case study of a backswamp in Southern Laos. *Society and Natural Resources, 18*, 41–60.
Unpatti, Pusdi-P S L. (1995). *Kajian hukum tentang norma adat dalam perlindungan lingkungan*. Ambon: Kantor Menteri Negara Lingkungan Hidup RI dengan Pusdi-PSL, Universitas Pattimura.
Veitayaki, J. (2001). Customary marine tenure and the empowerment of resource owners in Fiji. In G. R. South, G. Cleave, & P. A. Skelton (Eds.), *Oceans in the new millennium: Challenges and opportunities for the islands* (pp. 151–61). Bucharest: Dada.
Veloro, C. (1992). Complementary adaptations, contrasting images: Fishing and agriculture in a peasant village. *Yakara, 19*, 76–110.
Wahyono, A. (Ed.). (2000). *Hak ulayat laut*. Yogyakarta: Media Pressindo.
Walters, C. J. (1986). *Adaptive management of renewable resources*. New York: McMillan.
Walters, C. (1997) Challenges in adaptive management of riparian and coastal ecosystems, *Conservation Ecology, 1*(2). http://www.consecol.org/vol1/iss2/art1/.
Walters, C., & Holling, C. S. (1990). Large-scale management experiments and learning by doing. *Ecology, 71*, 2060–2068.
Zerner, C. (1989a). *Justice on the Makassar Strait: Criminal law, civil law and customary resource management in the Mandar Roppong fishery*. Jakarta: Central Fisheries Research Institute, Ministry of Agriculture.
Zerner, C. (1989b). Marine tenure in the Makassar Strait: The Mandar raft fishery. Paper presented at the 1st Annual Meeting of the International Association for the Study of Common Property, Durham, North Carolina.
Zerner, C. (1991a). Sharing the catch in Mandar: Changes in an Indonesian raft fishery (1970–1989). In J. J. Poggie & R. B. Pollnac (Eds.), *Small-scale fishery development - sociocultural perspectives* (pp. 42–70). Kingston, RI: International Center for Marine Resource Development, University of Rhode Island.

Zerner, C. (1991b, Sept). Reefs in a riptide: Community management of coastal resources in the Maluku Islands, Indonesia. Paper presented at the 2nd Annual Meeting of the International Association for the Study of Common Property, Winnipeg.

Zerner, C. (1991c, Sept). Imagining the common law in Maluku: Of men, mollusks, and the marine environment. Paper presented at the 2nd annual meeting of the International Association for the Study of Common Property, Winnipeg.

Zerner, C. (1991d). Community management of marine resources in the Maluku Islands. Jakarta: Fisheries Research and Development Project: (ms.).

Zerner, C. (1994). Through a green lens: the construction of customary environmental law and community in Indonesia's Maluku islands. *Law & Society Review, 28*(5), 1079–1122.

Zerner, C. (n.d.). The Pulau Ambo fishery: Resource royalties in the Makassar Strait (ms.).

Chapter 2
Pre-existing Fisheries Management Systems in Indonesia, Focusing on Lombok and Maluku

Arif Satria and Dedi S. Adhuri

Abstract In Indonesia pre-existing systems of fisheries management were delegitimized during the 'New Order Era' (1966–1998), and revived after the 'Reform Era' began, in 1998. Three such systems are examined; the *awig–awig* and *sawen* of North Lombok, and the *petuanan* and *sasi* of Maluku. Based on the pre-existing system that contained sawen, with its basic values and norms for integrated management of forest, farmland and coastal resources, local people took the initiative to revive three awig–awig, and adapted them to both combat destructive fishing practices and implement sustainable fisheries management. Sea tenure in Maluku is based on the concept of *petuanan laut*, the sea territory of a particular social group, to which 'the right to eat' (compounded from the rights of access, usage and exploitation) and 'the right of ownership' are attached. Sasi refers to the beliefs, rules and rituals regarding temporal prohibitions for a petuanan laut. The performance of pre-existing fisheries management systems is evaluated and national policy for them examined.

Keywords Awig–awig • Marine resources • Petuanan • Sasi • Sawen

2.1 Introduction

Pre-existing management systems have been retained in parts of Indonesia, and particularly in Sulawesi, Maluku and Irian Jaya. In Maluku Province sasi has continued since the seventeenth century, and refers to local communities' regulations that govern the harvesting of resources (Naamin and Badrudin 1992). It has the three

A. Satria (✉)
Jalan Bambu Raya 30, Taman Yasmin Sektor 7, Bogor, Jawa Barat, Indonesia
e-mail: arifsatria@ipb.ac.id; arifsatria@gmail.com

D.S. Adhuri
The WorldFish Center, Jalan Batu Maung, Batu Maung, 11960 Bayan Lepas, Penang, Malaysia
e-mail: dediadhuri@hotmail.com

K. Ruddle and A. Satria (eds.), *Managing Coastal and Inland Waters: Pre-existing Aquatic Management Systems in Southeast Asia*, DOI 10.1007/978-90-481-9555-8_2,

fundamental objectives of ensuring (1) fair and equal opportunities of access for community members in exercising their mutual rights to the nearshore fishery; (2) effective and sustainable management of sedentary marine species in nearshore waters; and (3) that community members can satisfy their subsistence needs and obtain an income from the community's marine waters. Sasi regulations are comprehensive, but focus mainly on the timing of the fishing season, regulation of target species and gear, and sanctions. Gear regulation also aims to promote an equitable distribution of income, in addition to attempting to manage fisheries sustainably. As a result, the use of gear like the purse seine is forbidden, as is diving apparatus to harvest Topshell, pearl oysters and other valuable aquatic resources. Sanctions may include fines, public shaming, and either temporary or permanent confiscation of fishing gear.

In North Sulawesi a pre-existing fisheries management system known as *seke* has existed since the Dutch colonial era (1521–1945). Rooted in Para Village, this system governs three types of fishing grounds: (a) *Sanghe*, a particular coral reef area that supports many fish species; (b) *Elie*, a fishing ground furthest from shore; and (c) *Inahe*, a border area separating the Sanghe and *Elie* (Wahyono et al. 2000). The term *seke* denotes a group of fishermen legally recognized by the village government who use a traditional fishing gear to catch scad (Wahyono et al. 2000).

To avoid conflict and ensure equity of access among the six groups of *seke* fishers in Para Village there is an agreed schedule that determines when and where a group can fish in the four suitable fishing areas. A payment of five to ten sacks of cement is imposed on whoever violates the agreement (Wahyono et al. 2000).

The *rompong* is an old established form of marine tenure in South Sulawesi that originated in the Bugis community at Makassar. It is practiced especially in the Makassar Strait, Bone Bay and Flores Sea. Satria et al. (2002) describe the rompong as providing fishing rights to areas of about one hectare and delimited by *adat* (customary law). Basically, rompong refers to a traditional fish aggregating device (FAD) made of bamboo poles and coconut fronds. A group of fishermen usually works together to construct a rompong. The area where they are placed is claimed as a property right, so that nobody can fish there except the rompong owner and rod-and-line fishers. This fishing right is usually obtained through transferability (in terms of legacy or granting) and/or is simply recognized by the community. Several rules apply for operating the rompong (Satria et al. 2002). Its owners have an exclusive fishing right, but must allow other fishers both unhindered transit and to fish with rod-and-line within the area. This property right may be transferred to other fishermen in the community. An owner not operating his romping must allow others making a request to fish in his area. Those who violate a rompong right usually have their boats sunk and nets burned by the right holder.

Marine tenure in Irian Jaya is based on village customary law, with the boundaries of a tenured area normally marked by natural features. Imaginary boundaries may also be included that extend the area to the horizon. However, those boundaries have recently become blurred owing to amalgamation of tribes and other social processes. Authority for fishing is strictly divided among tribes, with the larger being the more powerful. Matters related to marine territorial affairs are under *suku*

Sanyi authority, and issues regarding fishing technology are handled by *suku Drunyi*. Recently, authority has been gradually granted to smaller tribes, called *keret*, and the village often has assumed the traditional authority to protect its fishers from non-residents. Wahyono et al. (2000) describe the principal fishing rules, all of which were made at meetings led by the village unit (*Ondoafi*) and involving keret and adat members. Both church and village office representatives sometimes were involved, although they performed only advisory roles.

Important rules deal with fishing by outsiders, and the scheduling of fishing. Any outsider wishing to fish in a tribal area must first request through the village chief permission from the village unit, which consults with the property rights owning small tribes. That is followed by a meeting of a customary board (*dewan adat*) composed of the three main elements (village chief, church and tribal marine affairs leader). Local fishers must do the same when wishing to use modern gear. After having been issued a permit, a recipient is obligated to share his catch with the *adat* board. A series of sanctions may be imposed on violators including (a) an oral warning, (b) the confiscation of coconuts, (c) being ordered to hunt for pigs for customary ceremonies, and (d) being sentenced to death. All have been enforced at some time or other. Punishment for outsiders is different. Local fishers punish outsiders corporally for fishing illegally and by confiscating fishing gears and imposing a monetary fine. However, corporal punishment has declined under the influence of Christianity. Other rules concern the scheduling of fishing according to area of residence.

In Tobati and Enngros villages the marine tenure system is characterized by gender-based rules, with women being granted specific areas, especially in mangroves and shallow waters, to catch shrimp and crabs, and collect mollusks. These areas also become special places for women's education prior to marriage, and men are forbidden to fish in them (Wahyono et al. 2000).

A specific rule relates to the customary ceremony of *pele karang*, intended to invite fish. This is conducted for about 6–12 months, and usually at a village border with abundant coral. During the ceremony nobody may enter or transit the area, and violators are punished by *hobatan* (murder using magic). In recent years this ceremony has declined, under the influence of Christianity.

In the remainder of this chapter we examine in detail pre-existing fisheries management in Indonesia, based on the awig–awig (lit. 'a local rule')[1] of Lombok Barat (Fig. 2.1), and the petuanan and sasi of Maluku (Fig. 2.2). The awig–awig, a local institution that since the 1940s had managed resources effectively (Satria 2007a), exemplifies a revitalized management system. Those in Maluku represent the continuity of pre-existing systems.

In Lombok Barat, the revitalization of awig–awig was the local fishers' response to the national reform movement that began in 1998, with the dismissal of former President Soeharto, and which marked the end of the 'New Order Regime' (1966–1998) and the

[1]Although awig–awig was introduced when the island formed part of the Balinese Empire, the institution has long been an integral part of the cultural system of Lombok (Bachtiar 2002).

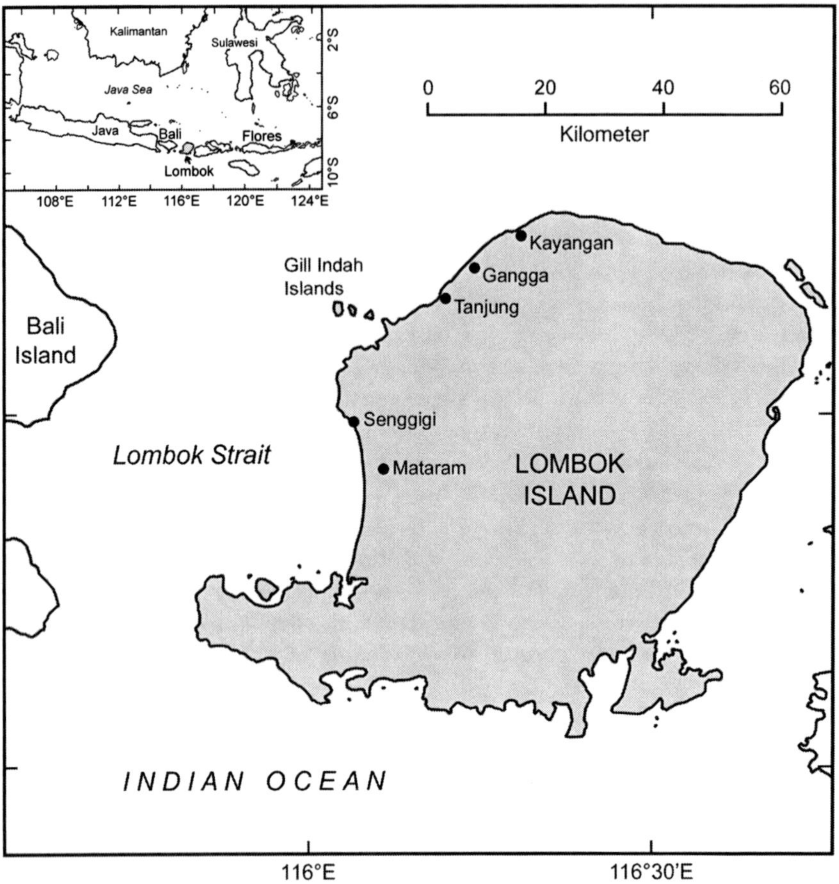

Fig. 2.1 Locations in Lombok Island

beginning of the 'Reform Era' (from 1998 until the present). This was a critical period during which political instability led to a lack of government accountability and authority to enforce formal rules in marine fisheries. The legacy of low enforcement rates was exacerbated by the Reform Era, which in effect created 'stateless areas' throughout much of Indonesia. Local people took advantage of that political vacuum to assume a new role as 'regulators'. The phenomenon of self-regulation in marine fisheries enabled them to replace various formal rules by revitalizing their own pre-existing institutions. As a consequence, following the recognition of increasing use of destructive fishing practices, especially blast fishing, in Lombok Barat local people decided to replace the formal rules for fisheries by revitalizing awig–awig (Satria and Matsuda 2004a).[2]

[2] Blast fishing was introduced to Lombok by Japanese soldiers, who fished with explosives during the military occupation of Gili, Lombok Barat, which began in 1942 (Satria and Matsuda 2004a).

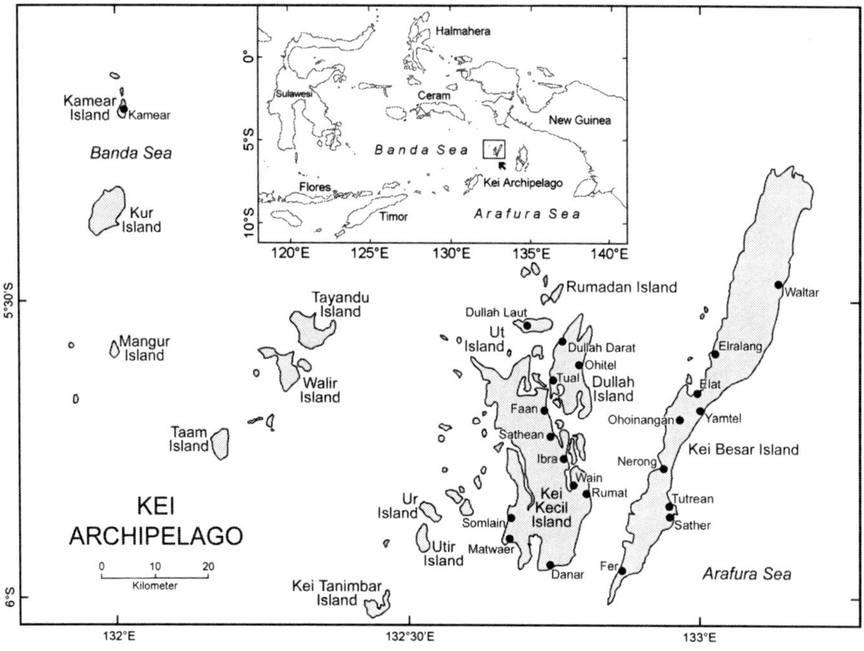

Fig. 2.2 The Kei Islands of Maluku Province

Since the mid-1980s, Maluku has been the geographical focus of attention on pre-existing marine resource management in Indonesia. And in Maluku the thematic focus has been on sasi, a system of beliefs, rules and rituals pertaining to temporal prohibitions on using a particular sea territory or specified resources within it. That is unfortunate, since it has diverted attention from petuanan laut, which is arguably a more important aspect in marine resource management, since it deals with property rights or marine tenure. In practical terms the issue of petuanan laut is important in Maluku, because whereas the practice of sasi was delegitimized and therefore weakened during the 'New Order Regime' (1966–1998), the tenurial practice of petuanan laut was strengthened (Adhuri 2002a).

How the awig–awig, petuanan and sasi operate is the focus of the first section of this chapter. We then examine the institutional performance of the awig–awig and sasi, and the impact on them of national policies. We conclude that pre-existing systems have an important role to play in the future management of small-scale fisheries and fishing communities throughout Indonesia. However, the systems analyzed in this chapter need further refinement and adaptation before they can both function properly in their own 'native' environments and serve as models for wider application.

2.2 The Awig–Awig of North Lombok

2.2.1 *The Sawen System*

Awig–awig is rooted in a pre-existing conceptual order known as sawen or *nyawen* (lit. 'boundary delineation') that prior to beginning of the 'New Order Regime' in 1966 was applied to forests, farmland and the coast. According to sawen, the forest is regarded as 'the mother' (*buana alit*), since it is seen as the source of water. Therefore if the forest is disturbed adverse effects would cascade through the entire ecosystem, via the hydrological system, to impinge eventually on farmland and the sea. Farming and fishing communities would be endangered through a decline in downstream agricultural resources, like irrigation water, which in turn would threaten coastal resources. This sophisticated human ecosystem concept provides the rationale for integrated resource management by the *mangku* authority (Satria 2007a).

Each section of the longitudinal profile has its own management authority, with distinct roles and responsibilities for resource sustainability. The forest is managed by the *mangku alas*, the *mangku bumi* manages farmland, and the *mangku laut* is responsible for marine resources. Reflecting the human ecosystem on which their responsibilities rest, these mangkus shared a strong commitment to managing resources in an integrated manner. As a result, coordination and collaboration among them was given a high priority, resulting in a functional interdependence of their roles.

Further, the mangku was a hereditary resource management authority that could be held only by a descendant of a mangku family. In other words, the status of mangku was ascribed rather than achieved. This arose from the belief that mangku families have both supernatural power and the knowledge to deal with resource management issues. The villagers' respect for the power of a mangku legitimates the mangku, and ensures voluntary compliance, as decisions of each mangku were perceived as a contribution to a safe and peaceful life.

A mangku had two main roles. First they had to maintain the traditional value of social and human–nature relationships that would ensure a harmonious community life. Second was resource management, which required a mangku to undertake *menjango* (survey or observation), *membanggar* (visual mapping and boundary marking and *membuka* (opening) (Kamardi 1999). Applied to forest, farmland and marine resources, these practices were based on a combination of traditional knowledge and myth. Many of the religious ceremonies that preceded them demonstrate that myths were influential in resource management.

These roles of mangku were based on clear concepts of resource management, despite being the result of a combination of traditional knowledge and myth. For example, in the management of marine fisheries sawen is identical with a seasonal closure concept. After observing conditions of the sea, the mangku laut would decide whether a fishing season should be closed. To initiate closure he installed two bamboo posts approximately 1.5 km distant from the shoreline, to mark the boundary of the closed area. The closed season, usually lasting about a month, was intended to lure fish close inshore, so that they could be easily caught during the following open season. Unable to operate in more distant waters, the fishers depended on the nearshore area.

In this sense, sawen can be interpreted as a way of dealing with a scarcity of fish in nearshore waters.

These underlying myths of *nyawen* were based on such scientific rationales as a closed season to enhance fish stocks, although explanations were usually given in easily understood normative terms, like prohibiting or allowing such activities via a taboo (*pamali*). After a sawen was issued, rules like prohibiting fishing in a particular area during the sawen period were established. Violators were sanctioned morally, in the most severe cases with social ostracism. Such rules were easily enforced because fishers regarded them as sacred. When the closed season ended, mangku officially opened the fishing season with a religious ceremony (*syukuran*), at which the fish caught on the first day were offered to the supernatural powers.

To enforce these rules and practices, mangku laut appointed *lang–lang* (traditional coast guards) to monitor and control each sawen. Since the position was voluntary and open to all fishers, most had experience of the job. When a violation was committed, a lang–lang had both to warn the violator and report the incident to the mangku laut, who would decide on an appropriate sanction. Most sanctions were moral, and designed to shame violators publically.

2.2.2 Awig–Awig: Revitalization of Sawen

Local people took advantage of the political vacuum during the reform momentum of 1998 to assume a new role as regulators, and to replace various formal rules by revitalizing their own pre-existing institutions. In North Lombok, revitalization of sawen resulted in the awig–awig. Four types were established in Kecamatan (Sub-District) Tanjung, Gangga, Pemenang, Bayan, and Kayangan (Satria and Matsuda 2004b) (Table 2.1).

2.2.2.1 The Protection of Marine Fisheries Resources: Fishers' Council of Northern Lombok

The Awig–awig *Lembaga Masyarakat Nelayan Lombok Utara* (*LMNLU*) or 'Fishers' Council of Northern Lombok') was established in March 2000 by the fishers of the three *kecamatan of* Tanjung, Gangga and Pemenang to prevent such destructive practices as blast fishing and the use of poisons. *Samudera*, an NGO in Lombok Barat, participated in the establishment of this awig–awig. It functions as a lead organization of fishers in Northern part of Lombok Barat (Lombok Utara) that coordinates the awig–awig of each village. The village chiefs, sub-district chiefs and an NGO witnessed the promulgation of the awig–awig.

The rules devised for blast fishing and the use of poisons are that those fishing in this way will be taken to the official authority to sign a statement promising that they would not repeat the offense, and to pay a fine equivalent to USD 977.[3] Should they

[3]The currency rate has been converted at IDR (Indonesian Rupiah) 10,235.21 = 1 USD (July 07, 2009).

Table 2.1 Awig–awig system in Lombok Barat (2000–present)

Type	Rules	Sanctions	Compatibility with formal laws
Awig–Awig Gili Indah in Kecamatan Pemenang	a. Zoning system b. Prohibition of destructive fishing practices c. The mechanism of authorization for appropriation activities	Fine, and damaging seaweed culture	a. The Fisheries Law No. 9/1985: • Fine equivalent to USD 2,442 • Confinement of 6 months to 10 years c. The Environmental Law No. 23/1997: • Confinement of 10–15 years • Fine equivalent to 48,851–73,276 USD
Awig–Awig Kelompok Nelayan Pantura in Kecamatan Kayangan	a. Prohibition of fishing by blasting, trawling, and gill netting in awig–awig area b. Closed season	Fine, and confiscating fishing gear	a. Fisheries Law No. 9/1985: • Fine equivalent to USD 2,442 • Confinement of 6 months to 10 years b. Environmental Law No. 23/1997: • Confinement of 10–15 years • Fine equivalent to 48,851–73,276 USD c. Provincial Regulation of NTB No. 5 /1996 • Fine equivalent to 4.9 USD • Confinement of 6 months
Awig–Awig Sari Laut in Kecamatan Bayan	Prohibition of fishing by dynamite, potassium cyanide, trawl net	Fine, and physical sanction without resulting in death	a. The Fisheries Law No. 9/1985: • Fine equivalent to USD 2,442 • Confinement of 6 months to 10 years b. The Environmental Law No. 23/1997: • Confinement of 10–15 years • Fine equivalent to 48,851–73,276 USD c. The Provincial Regulation of NTB No. 5 /1996 • Fine equivalent to 4.9 USD • Confinement of 6 months

(continued)

Table 2.1 (continued)

Type	Rules	Sanctions	Compatibility with formal laws
Awig–awig LMNLU in Kecamatan Tanjung, Pemenang, Kayangan, and Bayan	Prohibition of fishing with dynamite and potassium cyanide	Fine, physical sanction without resulting in death, and burning gear and boat	a. The Fisheries Law No. 9/1985: • Fine equivalent to USD 2,442 • Confinement of 6 months to 10 years b. The Environmental Law No. 23/1997: • Confinement of 10–15 years • Fine equivalent to 48,851–73,276 USD

This table is based on ideas presented in a speech by the Chief of the Marine and Fisheries Service Office of Lombok Barat, in 2002

continue to use such fishing methods, their fishing gear and boat would be burned by the local fishers. Finally, if despite those actions a fisher persists in using destructive fishing methods, he will be punished corporally but not killed by the local people.

Apart from the rules regarding corporal punishment, those devised by the local fishers are compatible with the formal rules of the government (Table 2.1). Because the local government regards the local rules as more effective than formal rules in preventing destructive fishing, it has neither challenged the authority of awig–awig nor advised that the sanctions be withdrawn.

An Executive Committee elected by the fishers has the implementing authority. The organizational structure of the *LMNLU* consists a Board of Advisors, composed of the officials of the Sub-Districts of Pamenang, Tanjung and Gangga, and the village chiefs of Pamenang, Tanjung and Gondang, and an Executive Committee consisting of a Chairman, Vice-Chairman, Secretary, Treasurer, and Bureaux (for sea security, beach cleaning, social welfare, conservation and rehabilitation).

The highest authority is vested in the General Assembly, held every 3 years and open to all fishers of Lombok Utara. The General Assembly elects the Executive Committee and formulates the programs of the LMNLU. Because of the large geographical area involved, the chairman's role is to coordinate the awig–awig of each village in Lombok Utara. The role of the Bureau of Sea Security is monitoring fishing activities and arresting those who violate the rules. The task of the Bureau of Beach Cleaning is to enhance awareness of environmental sanitation and management of fishing boat anchorages.

The awig–awig *Kelompok Nelayan Pantura* in Kayangan Sub-district. was revitalized in August 2002 by local fishers acting alone. Aimed at protecting marine fisheries resources, the rules prohibit blast fishing, trawling and use of gillnets (*seret* net). All fishing is prohibited when a sawen or closed area has been declared, and sanctions are imposed on violators. Those catching ornamental fish are fined the equivalent of USD 49; those blast fishing are fined the equivalent of USD 489, and their boats and gear confiscated (Photo 2.1); those either trawling or using a *muroami* (drift-in net)

Photo 2.1 A violator's fishing boat confiscated by villagers at Gili Air, Lombok Barat, Indonesia

are fined the equivalent of USD 1,465, and their boats and gears confiscated; and those fishing with potassium cyanide are fined the equivalent of USD 244.

2.2.2.2 Prevention of Destructive Fishing Practices: The Awig–Awig *Sari Laut*, Bayan Sub-District

This awig–awig was established by local fishers in October 2000, to prohibit blast fishing, the use of potassium cyanide and trawling. The *Sari Laut* NGO and village government are supporting and advisory bodies, and the lang–lang laut play a major role by monitoring implementation by warning violators, making them to promise not to repeat the offense, and confiscating their boats. Should a violator persist, the *Persatuan Nelayan Sari Laut* (the local fishers' organization) and lang–lang laut can arrest them, confiscate their gear and impose a fine. The following fines are specified; for blast fishing the equivalent of USD 684, for using potassium cyanide the equivalent of USD 977; and for trawling the equivalent of USD 489. After a third violation, *Persatuan Nelayan Sari Laut* and *lang–lang laut* and other fishers first will punish the violators corporally, and then hand them over to the police.

2.2.2.3 Coral Reef Management and Prohibiting Destructive Fishing

The awig–awig at Gili Indah Village was established in 1999 to manage coral reef conservation by zoning for tourism and fisheries, and to prohibit destructive fishing practices. It includes three kinds of rules: (a) Those to establish protective, buffer and exploitation zones; (b) those to separate permitted and prohibited activities by zone; and (c) those to authorize appropriation activities. Zones were established considering the condition of the coral reefs. Where coral was plentiful protection zones were restricted, and only snorkeling and diving allowed, and net fishing and seaweed culture forbidden. Diving, snorkeling and angling are permitted in the buffer zones, whereas in the exploitation zones most activities were allowed, except drift-in and gill netting. Collecting marine biota (including turtles, turtle eggs and the giant clam [*Tridacna gigas*]) except fishes, whether for commercial purpose or private, is prohibited in all zones (article 20), although collecting marine biota for scientific purposes must be licensed (article 21). Pearl culture is prohibited in all zones within 50 m of the outer reef slope (article 22), and seaweed culture must be authorized by the sub-village (*Dusun*) chief (article 23).

The three awig–awig described here are initiatives taken by local people to overcome destructive fishing practices. They were aware that awig–awig were part of a local pre-existing management system that was delegitimized after 1966, when the 'New Order Period' began. However, they were also aware that the pre-existing system contained sawen, with its basic values and norms for resources management. Sawen was revitalized and adapted to contemporary conditions.

2.3 The Maluku Case

From the latter half of the 1980s discourse on pre-existing marine resource management in Indonesia concentrated on sasi, and neglected property rights or marine tenure, which in Maluku is known locally as petuanan laut. Because of that we begin the discussion of the pre-existing system of marine resource management in Maluku with an examination of petuanan laut (Table 2.2).

Table 2.2 Some basic characteristics of Petuanan and Sasi

Type	Rules	Sanctions	Note
Petuanan	a. Boundary definition	Driven away	Petuanan practice is an integral part of the social construction of society. During conflict, petuanan becomes part of the conflicted issue
	b. Beyond subsistence use, exlusive use/ exploitation of marine territory for right-holding unit members only	Monetary fine and/or traditional goods	
Sasi	a. Closed season	Monetary fine	The closed period shortened from 3 years to 1 year
	b. Gear restriction	Confiscation of catch	Transfer of control from community to village government
	c. Size limit for *Trochus niloticus*		

2.3.1 Petuanan Laut

The basis of tenure practice in Maluku is embodied in the concept of petuanan,[4] which is generally understood throughout the region as the estate or territory of a particular traditional social group (Zerner 1992). The concept includes both land and sea, a linked expression found in the pairing of such terms as petuanan laut (sea estate), *met* (coastal area) and *roa* (sea), referring to a sea territory on the one hand, with *petuanan darat* (land estate), *nuhu* (island), and *nangan* (land) referring to a land territory on the other. For sea territory the object of the ownership is called 'petuanan laut'.

The conception of petuanan laut boundaries varies among communities. In a seaward direction some claim that petuanan laut includes the area from the maximum high tide line to where shallow water meets deep sea (*tohar*). Others believe that the seaward boundary of petuanan laut is 'as far as eyes can see,' whereas yet others associate the seaward boundary with technology, claiming that their territory includes the entire area in which their boats and gear can operate. Rahail (1995), the late 'king' of Maur Ohoi Wut, in Watlar, Kei Besar Island, claimed that the petuanan laut of his domain covered the area as far as *tahait ni wear*, meaning the water more than 10 km from the beach and more than 5,000 m deep.

On land a petuanan laut boundary always is associated with the land boundary that divides two traditional domains. This mostly is a natural landmark, such as a rock, hill, embayment, or large tree. Although an easily visible and named natural landmark, a land boundary is often a source of conflict between neighboring communities. This occurs because boundary claims, like the entire petuanan area, are legitimated by oral history, of which there are often multiple versions open to multiple interpretations.

Traditionally, two rights are attached to the territory of petuanan laut. The first is *hak makan* ('the right to eat'), which is compounded from the rights of access and usage. Fishing operations provide an example of how hak makan is exercised. The second right is *hak milik* ('the right of ownership'). Hak milik is superior to hak makan; not only may holders of this right of ownership freely use the territory (hak makan), but they can also transfer their hak makan to another party. A contract between a representative of village leaders under the leadership of the village head and a fishing company concerning permission for the latter to fish in the village sea territory would exemplify how hak milik holders transfer their hak makan to another party.

These two rights are not distributed equally within a community. Whereas every member of a community has 'the right to eat', 'the right of ownership' is held only by descendents of the originating kin groups, whose ancestors founded the community, as recorded by oral history. Therefore whereas all community members

---

[4]*Petuanan* is derived from *tuan*, lit. 'owner' or 'master'. The prefix *pe* and suffix *an* add the notion of place to 'tuan'.

can participate in every activity to exploit a petuanan laut, only members descended from the originating kin groups can transfer the use rights to a second party. Such a transfer, either through auction or by a contract, is usually decided by a representative of the originating kin groups.

Based on those concepts, a petuanan laut is an exclusive territory, the use of which is under the control of a community. Only the members of that community can use it freely. Outsiders seeking access, particularly for commercial purposes, must obtain permission from the originating kin groups. Only then can an outsider become involved in any commercial activity that exploits a petuanan laut.[5]

The terms 'community' or 'traditional domain' are used here to refer to a social unit claiming ownership of a particular petuanan laut. However, 'traditional domain' is applied to different sizes and types of social unit. In the Kei Islands, for example, various differently constituted traditional domains control petuanan. Some are attached to a settlement (*kampung*) community, such as *petuanan kampung* Hollay and Hoko, on Kei Besar Island. Despite being administered as a single village, each of these two settlements controls its own petuanan autonomously. Other petuanan are controlled by a *negeri* or *desa* (village) community. Dullah Laut Village provides an example. Although it consists of two different settlements, they share control of a single petuanan. As a result, Dullah Laut Village as a single unit deals with outsiders seeking access to the petuanan. A federation of villages that traditionally is considered to be a kingdom (*ratschap*) illustrates another traditional domain that controls a single petuanan. This is exemplified by Ratschap Ibra, on Kei Kecil Island, where the three villages of Ibra, Sathean and Ngabub control a single petuanan. In this case no one village autonomously handles petuanan issues, and all three together, under the leadership of King of Ibra, are entitled to speak for it. Finally, some petuanan were controlled by a larger social unit, such as moiety or an ethnic group.

Members of some coastal communities not only claim ownership of a sea territory based on the concept of petuanan laut, but also have developed sets of pre-existing rules that further address in detail the inter-related issues of who may use what resources contained in the territory, and when and how they are permitted to do so. Such sets of regulations are called sasi (lit. 'to witness' or 'witness').

2.3.2 Sasi

Sasi refers to a system of beliefs, rules and rituals pertaining to temporal prohibitions on using a particular resource or territory. When sasi is applied (*tutup*) to a particular resource, no usage whatsoever is permitted until the sasi is lifted (*dibuka*).

[5] In contrast, outsiders do not require permission for non-destructive subsistence activities. However, community members will observe outsiders to evaluate their activities, and would not hesitate to drive them away should they suspect that their activities are illicit.

When applied to a coconut palm, for example, nuts may neither be harvested nor fallen ones used. The prohibition is applied to everybody, including the owner of the resource.

The various types of sasi are differentiated by the resource or territory concerned, as well as by the belief system, type of ritual leadership, and location of the rituals.[6] Common examples of sasi applied to resources are *sasi kelapa*, for coconut, and *sasi lola*, for Topshell (*Trochus niloticus*) and other shellfish. Territorial sasi is differentiated into that on land (*sasi darat*) and at sea (*sasi laut*). Several types of sasi are distinguished by belief system, ritual leadership and location. *Sasi negeri* (village sasi) is based on local belief, with the rituals of applying and lifting sasi led by traditional leaders, and performed at sacred places in the village. *Sasi gereja* (church sasi) is based on Christianity, with rituals conducted in a church by a priest, according to Christian beliefs. Similarly, *sasi mesjid* (mosque sasi) is based on Islamic belief, with rituals led by an *imam* and conducted in a mosque.

Only by using the term *sasi laut*, often called *sasi meti* or *sasi labuhan*[7] or *sasi bia lola*, is specific reference made to pre-existing marine resource management. Sasi laut is applied to either an entire petuanan laut, or to just a portion of it. In the ritual of applying sasi (*tutup sasi*), the leader announces the sea boundaries of the area under sasi, and the resources thus regulated. He announces the gear types and fishing techniques excluded from sasi regulations, and, in the case of sasi negeri, stipulates a fine for violators. The fine can be a sum of money or traditional goods like antique gongs and cannons. Confiscation of the gear, catch or other items used in the illegal operation is also a common action following an apprehension.

The same ritual practitioner performs *buka sasi*, a ritual to open or lift the sasi regulations. In addition to communicating with the spirit world, the ritual also functions to inform about the conduct of harvesting. The information provided usually includes the resources and quantities that may be harvested, participation, permitted gear types, the manner of distribution, and the length of harvesting period.

Sasi bia lola, applied to Topshell, is one of the commonest forms of sasi used in Maluku. No harvesting is permitted when the sasi is in operation, and in some places diving or fishing using gear considered to disturb either the Topshell or its habitat is forbidden. In former times a sasi would be closed for three or more years, but since the 1980s, frequently has been opened annually, with the harvest period ranging from few days to two weeks.

Regulations pertaining to participation and allowed gear and fishing techniques differ among communities. At least since 1968 in Nolloth village, Central Maluku, only appointed people could participate in harvesting, whereas in communities on the east coast of Kei Besar Island, in Southeastern Maluku, representatives of all households could dive for Topshell. Gear was limited to just diving goggles, and

---

[6] See Monk et al. (1997) for a more detailed account.

[7] The second word in each pair refers to local names of locally controlled sea territory.

free diving was the only technique allowed. The generally accepted minimum harvestable size was a diameter of 'three fingers', or approximately 6 cm.

The distribution of the catch also varies among communities. During the 1990s in Nolloth, for example, all divers were hired by the village government. They were either paid a fixed amount of cash, or with a percentage of the value of the total catch. In Watlar village, on Kei Besar Island, the divers received the meat and 20–30% of the total sale price of the shells. In both villages, the balance of the sale price was supposed to be used for community infrastructure projects. In Nolloth some of the income was used to pay for the special traditional committee that oversaw the implementation of sasi.

The sasi laut and sasi for Topshell is administered by either a special traditional committee (*kewang;* lit. 'police'), or by the traditional government. A kewang consists of a leader (*kepala kewang*), a secretary and some functionaries. Among other duties, this committee is responsible for leading the implementation of sasi laut. This includes conducting both opening and closing rituals, monitoring the territory to ensure no rule violation, and sanctioning violators. In Central Maluku, the kewang together with the village head usually leads the practice of sasi laut. In Southeastern Maluku traditional government usually organized the practice. In Nolloth and Haruku villages, on Saparua and Haruku Islands, of Central Maluku, respectively, a kewang manages both land and sea petuanan. In contrast, communities in Southeastern Maluku lack a special committee to observe sasi, so all everything is handled by community government officials.

2.3.2.1 An Interpretation of Sasi

Since the early-1980s sasi has been interpreted and evaluated by various agencies and scholars. Initially, the discourse was aired widely by NGOs, research centres and legal scholars based in Ambon, Maluku. A research report prepared jointly by an NGO and academic researchers from the Law Faculty and Maluku Research Centre at the University of Pattimura observed that "[Sasi] strongly supports conservation of living marine resources ... in addition to being rather useful because it regulates the resource use, extraction and protection, it also ensures an even distribution of the harvest"[8] (translated from Anon 1991: x, see also Pusdi-PSL Unpatti 1995).

That is consistent with a definition by a *kewang* leader in Haruku village, Central Maluku (Kissya 1995), who notes that "sasi can be described as a prohibition on the harvesting of certain natural resources in an effort to protect the quality and population of that biological natural resource (animal or plant) " (Kissya 1995: 4). This argument is also supported by a legal scholar based in Ambon (Lokollo 1994), who went further and suggested that sasi should be considered as the basic model for the national policy on rural environmental management (Lokollo 1988).

[8] The translation is taken, with slight modification, from Zerner (1994: 1114).

Such arguments persist. However, a more critical perspective on sasi emerged from the early-1990s, based on the argument that earlier thinking was misleading because it was constructed without reference to the historical and socio-political context of sasi. Thus it was argued that "[S]asi has undergone considerable change over the past 400 years … it has developed from a ritual protection of communal resources to a governmentally regulated regime of agro-ecological control of private and common resources, and from there to a largely commercialized and privatized means of theft prevention." (Benda-Beckmann von et al. 1992: 5). Such historical analyses demonstrate that the practice of sasi has been mostly crafted by elites from inside and outside local communities.[9] In the late colonial era, for example, the ratification of sasi rules was initiated by local traditional elites in collaboration with local Dutch officials to meet the economic and political interests of both (Zerner 1994: 1087).

More recent elite initiative was exemplified during the 1960s by the sasi laut of Nolloth village, on Saparua Island (Zerner 1991). During the 1950s, the market demand for Topshell reached Maluku. Thus stimulated, the head of Nolloth village started raising the issue of a sasi for Topshell on the village sea territory. It was enforced in 1968. However, he made some changes to sasi practice. Before the sasi was implemented the sea territory was open to all villagers, who could benefit from harvesting Topshell. When the 'new' system of sasi was introduced the village headman declared that the territory was closed to community members, and the village administration took full control of it. Henceforth all income from the Topshell harvest would be for the village, and was earmarked for such village programs as roads and public toilets. Problems emerged regarding distribution of the income, and villagers began to question whether the money from the Topshell was really used to benefit the entire community. They also asked why the village committee hired outsiders to harvest the Topshell, when it should have hired villagers.

Studies on the contemporary practice of sasi provide further insights into the local realities. Pannell (1997: 297) notes that

> [T]he practices referred to and associated with *sasi* in the marine environment of Luang [south-eastern Maluku] minimally involve the interest and actions of residents of this island, the commercial machinations of regional traders and internationals exporters, the fashions and fads of distant consumers, the compliance and blessing of the Church and its agents, as well as the endorsement of village representatives of local government institutions and the support of government personnel from other jurisdictions. In addition, let us not forget those fishermen who, though their non-sanctioned exploitation of local marine resources, contribute to the social delimitation of the efficacy of invoking *sasi*.

Having noted the involvement of various agencies, as well as interests, in the practice of sasi, Pannell suggested that it might mean different things to different agencies with different interests. For example, "… for the traders the opening of *sasi*

[9] However, data on the pre-colonial context are very limited, making convincing arguments difficult to construct.

ensures that they enjoy exclusive rights of purchase [on the harvest] ... for people on Luang, the payments made by traders [for his {*sic*}monopolistic rights to buy the harvest] also amount to *de facto* recognition of their rights and interests as customary and communal title holders of these marine areas." (Pannell 1997: 296). In evaluating the contemporary sasi practice in Watlar village on Kei Besar Island, among other things it was found that the monopolistic control by a traditional leader in the village had stimulated villagers to both overharvest Topshell and question the distributional equity of the practice (Antunès and Dwiono 1998; Antunès 2000).

These historical and contemporary analyses raise questions about the conservation and equity factors that have been presented as an inherent part of sasi. When the discourse on sasi is analyzed in its socio-political context, it is evident that local traditional leaders, NGOs and scholars have been actively engaged in the process of 'greening' it. On this point, Zerner (1994) writes that the political context of the emergence of green sasi includes both a growing environmental awareness and also the resistance of local elites and NGOs to growing resource control by the central government and fishing industry. In this sense, the discourse of green sasi can be seen as a political discourse that aims to empower marginalized local people.

2.4 Institutional Performance

Here the performance of pre-existing fisheries management systems is evaluated using indicators modified from the six design principles of traditional fisheries management system proposed by Ruddle (1998). These are definition of territorial boundary, rules, rights, authority, monitoring and surveillance, and sanctions.

2.4.1 Clearly Defined Territorial Boundary

The territories of awig–awig *Kelompok Nelayan Pantura* (Pantura Fishers Group) and *Sari Laut* are clearly bounded, because the awig–awig area is similar to the sea area over which a village has jurisdiction. In addition, the awig–awig of Gili Indah Village has a clear territorial boundary located 30–100 m from the coastline around its island. In principle, the territory of the awig–awig is the water area in which the coral reefs are located. The territory is then divided into various zones, each with different usage and regulations. Zonal boundaries are delimited by such physical marks as bungalows, trees, floating balls, buoys, and other features.

In contrast, the territory of the *LMNLU* is not clearly distinguished, because it is not a territory-based organization, unlike the *Kelompok Nelayan Pantura* and *Sari Laut*. Although in practice the *LMNLU* is positioned as a coordinating organization to deal with destructive fishing practices, it was established by fishers in the different Sub-districts, who understand the importance of resource sustainability.

In Maluku, the boundaries of managed marine areas are physically distinguishable, since they are delimited by natural marks. However, exclusive claims to territories can be contested, because their source is a narrative relating to territorial origins. There sometimes exist multiple versions of a narrative, and these are open to various and conflicting interpretations.

2.4.2 Legitimacy and Enforceability of Rules

Because consensus building is conducted by local people, and therefore an awig–awig is regarded as being legitimate, its rules are easily enforced. This is particularly true for the *Kelompok Nelayan Pantura* and *Sari Laut*,

In contrast, the legitimacy of the *LMNLU* is not as strong, owing to its inherent characteristics and representational problem. The *LMNLU* covers many sub-districts, within each of which exist many fishers' groups with different interests that have not been organized into a fishers' association. As a result, the *LMNLU* deals with the problem of fishers' representation of each sub-district, so it is not legitimate in fishing communities lacking an awig–awig. These fishers assume that the *LMNLU* cannot represent the fishers of Lombok Utara. Nevertheless, the *LMNLU* is legitimate in the fishing communities where an awig–awig exists, because they share a mission to end destructive fishing practices.

However, the existence of an awig–awig does not necessarily mean that rules are easily enforceable. This is the situation at awig–awig Gili Indah, where, apart from banning the drive-in net, zoning and prohibition of blast fishing, most rules cannot be enforced. At Gili Indah there is a crisis of legitimacy within the community, and conflict among stakeholders is frequent.

In Maluku the legitimacy and enforceability of petuanan and sasi rules varies by location. Where legitimacy is strong and the implementer well respected, the rules are obeyed by most people. Elsewhere conditions have weakened. Although the basic regulations are rooted in tradition, not all community members have always agreed with various adjustments and modifications to them. It was often the case that adjustments were made only on the initiative of traditional elite, to serve its own interest. Where that occurred community members who felt pushed aside resisted the new regulations, resulting in a decline in the pre-existing management practices.

2.4.3 Monitoring

The monitoring authority of an awig–awig is vested in the *pamswakarsa* (voluntary task force) of the LMNLU, and in the lang–lang laut of the *Kelompok Nelayan Pantura* and *Sari Laut.* Both are composed of local fishers. Monitoring activities are conducted intensively by both the *Kelompok Nelayan Pantura* and *Sari Laut*, unlike the LMNLU, because of their different mandates. There is a fixed monitoring schedule for each member, apart from the LMNLU, owing to its limited monitoring capacity and unclear territorial boundary. Therefore the LMNLU regards all fishers in Lombok Utara as monitors, and hopes they will call the pamswakarsa if violations occur.

The monitoring authority in the awig–awig Gili Indah is the *satgas* (security task force), which focuses on blast fishing. The satgas were appointed by businessmen in the tourist industry, who provide financial support for their operations. Therefore monitoring in awig–awig Gili Indah is not done on a voluntary basis.

In Maluku monitoring is conducted either by a special committee, called kewang, or by village officials. The kewang seems to perform better, probably because it has only to implement the petuanan and sasi, whereas village officials are concerned with general village management, and so can devote little time to monitoring.

2.4.4 Graduated Sanctions

Overall the regulations of the awig–awig have proven enforceable. This is especially true of the prohibitions on destructive fishing, as indicated by the decrease in blast fishing after the awig–awig were established, and the success in arresting violators. However, enforceability of sanctions also resulted from police and *KSDA* (Station for Natural Resources Conservation) support.

When awig–awig rules are violated, especially those regarding blast fishing and the use of poison, the LMNLU is invited to join the awig–awig authority to devise a sanction. LMNLU is supposed to be responsible for eradicating destructive fishing practices in Lombok Utara, even in areas where awig–awig exist.

In awig–awig Gili Indah sanctions for violation of the zoning rule ineffective, whereas those regarding blast fishing remain valid. Previously, the satgas of Gili Indah was firmly united, and the enforcement of sanctions was also supported of the police and KSDA, which had representatives in the popular tourist destination of Gili Trawangan (Satria et al. 2006).

In Maluku, various degrees of sanctions have been applied to petuanan and sasi. In Haruku, for example, rules are have been observed and, consequently, sanctions have been few, a situation attributable to powerful and committed kewang. In contrast, Antunès (2000) reported that the sasi in Kei Besar was not implemented well, and many people harvested undersized Topshell without fear of sanction.

2.4.5 Legitimate Authority

The awig–awig is linked with higher institutions, especially for the prohibition of blast and poison fishing. The LMNLU collaborates well with KSDA, Dinas (Local Fisheries Service) and an NGO, whereas the Kelompok Nelayan Pantura is relatively exclusive and, apart from the LMNLU as a coordinating body, is not linked with other agencies. The Sari Laut is an NGO that supports the institutions with technical assistance, facilitation and advice. However, the legitimacy of the traditional authorities in enforcing awig–awig rules is relatively high, both in the eyes of external parties and local people. The awig–awig Gili Indah is linked with a higher institution, especially regarding the prohibition of blast fishing. However, that authority tends to include only the tourist industry businessmen, and fishers are excluded. As a result, although the external parties regard it as legitimate, it is weak within the local community.

The legitimacy of petuanan and sasi was strong and, although not formally supported by or linked to either government regulations or institutions, it had informal local government support. In Maluku, pre-existing management practices, authorities and institutions are often stronger than the government. However, in communities where traditional leaders do not perform well or are either proven or accused of manipulating tradition for their own interest, people question the leadership and even the tradition. In this circumstance the implementation of petuanan and sasi rests on an unstable foundation (Adhuri 2005).

2.5 National Policy on Pre-existing Fisheries Management

Pre-existing fisheries management was not recognized during 'The New Order Period' (1967–1998), based on *Undang–Undang* No 5 1979 ('The Rural Governance Law'), which required a uniform system of rural governance nationwide. Thus pre-existing systems were neglected and local people, having no responsibility for or participation in the management of marine resources lacked any sense of stewardship for conserving and protecting them. Under those conditions marine resources and became depleted.

The situation began to change in 1999, with the beginning of the 'Reform Era' (1999 to present), the establishment of the Ministry of Marine Affairs and Fisheries (MMAF), and passage of the Local Autonomy Law. At the beginning of the Reform period, Minister of Agriculture Decree No 392/1999 was issued as a revision of the Minister of Agriculture Decree No 607/KPTS/UM/9/1976 on fishing zonation. Three zones were fixed, as follows: (a) Zone I.a (0–3 nm) is reserved for traditional fishers using boats without engines, and Zone I.b. (3–6 nm) is reserved for traditional fishers with either outboard engines or using a boat of less than 5 gross tons; (b) Zone II (6–12 nm) is reserved for fishers using a boat of less than 60 gross tons, and (c) Zone II (6–12 nm) is reserved for fishers using a boat of less than 200 gross tons.

This regulation, aimed at protecting small-scale fishers, contains use rights instead of management rights. Nevertheless, the limitation of traditional fishers' rights to access and withdraw the resources only within Zone 1 ignores the possible existence of traditional fishing grounds seaward of Zone 1 (Saad 2003).

Using Ostrom's (1990) approach, Satria (2007) reviewed coastal and fisheries policy, emphasizing the protection of local people. The related formal laws are the revised Fisheries Law No 31/2004, the revised Local Government Law No 32/2004 (popularly called the Local Autonomy Law) and the Coastal and Small Island Management Law. Together they demonstrate a meaningful commitment to empower fishers and develop small-scale fisheries, because the government is responsible for providing financial support and promoting fisheries cooperatives. The revised Fisheries Law No 31/2004 appears supportive of pre-existing fisheries management systems, because in article 61 it addresses the access and withdrawal rights of the small-scale fishers. It states that "small-scale fishers are free to fish in all fisheries management areas of the Republic of Indonesia" (article 61). This article was inspired by the Local Autonomy Law No 22/1999, elucidation of article 10, and its revised version No 32/2004, and elucidation of article 18. By the latter "small-scale fishers are defined as traditional fishers who engage in fishing using traditional fishing technology and on whom an enterprise certificate and tax are not imposed, and are free to fish in all fisheries management areas of the Republic of Indonesia". This means that small-scale fishers gained rights to access and withdraw marine resources in all areas.

There are two critical issues regarding fishing rights as stated within the revised Fisheries Law and the revised Local Autonomy Law in the Reform Period (Satria 2007b). The first is that the articles addressing fishing rights for small-scale fishers ignore pre-existing property rights. Generally, many fishing communities develop property rights based on either their own local rules or customary law. These pre-existing rules address management rights by which fishers manage some marine resources and exclude outsiders seeking to fish in designated areas. As a consequence of limited communication conflicts will arise if all small-scale fishers can fish freely without prerequisites, since they may be either unaware of or unwilling to accept the local operational rules devised by the local fishers.

The second issue is that, although the revised Fisheries Law No 31/2004 is better than Fisheries Law 9/1985, there is no article in it that explicitly addresses management rights, although they have existed for centuries. This means that the local fishers must follow the rules devised formally from outside, either by the central or a local government. The critical issue is if the formal rules do not coincide to some degree with social norms, or are perceived as being unfair, they provide an immediate incentive for violation. Eventually, the rules are likely to be only weakly enforceable, resulting in poorly managed marine fisheries resources.

However, by the Local Government Law 22/1999 the central government must transfer the authority for marine resources management to local governments. Based on a case study in Lombok Barat (Satria and Matsuda 2004b) the positive impacts of that decentralization policy are state recognition and strengthening of pre-existing fisheries management systems, and devolution of fisheries management to local people. These results demonstrate that decentralization can be an external factor for strengthening pre-existing fisheries management systems, and indicates that to some extent local autonomy indirectly affects their importance and strengthening.

The direct policy of recognizing pre-existing fisheries management is stated in the Coastal and Small Island Management Law No 27/2007, Article 62, which clarifies that communities and the private sector have an equal opportunity to participate in the planning, implementation and supervision of coastal and small islands management. It also mentioned in article 9 (3) that planning of zones is done considering the obligation to allocate community space and access in coastal and small islands.

That demonstrates that all coastal stakeholders are guaranteed fair treatment. Further, in article 61 it is affirmed that (1) the Government admits, respects and protects the rights of customary communities and traditional communities and local regulations of coastal areas and small islands that have been in operation for generations, and (2) it admits rights of customary and traditional communities and local regulations as a reference of coastal and small island management.

The Agrarian Principle Law (UPL) of 1960 also contained an article, stated in general terms, about the admission of customary rights. Also, in article 16 subsection 2 the UPL mentioned conservation and fishing rights. But this was of little importance because it barely regulated a withdrawal right, and not a management right, admission of which is of fundamental importance in the devolution of coastal management. However, the policy remains to be implemented via either a Government Act (*Peraturan Pemerintah*) or a Ministerial Decree (*Peraturan Menteri*).

2.6 Conclusions

The awig–awig, petuanan laut and sasi contain elements essential for the development of workable fisheries resource management for modern conditions. These include communal marine tenure and a combination of such input and output controls as seasonal closure, gear limitation and target size restrictions. These are all modern instruments of management that often cannot be implemented owing to the resistance of fisheries stakeholders, among other impediments. Further, as the practice of sasi demonstrates, some level of community compliance is fundamental to the successful implementation of these instruments.

In Indonesia pre-existing marine resource management systems can play an invaluable role in the protection of small-scale fishers in modern society. The prohibition of trawling, drive-in nets and other larger-scale fisheries in Lombok and Maluku assures exclusive access rights for local traditional fishers. The indirect benefit of such rules is reduction of social conflict and a theoretical improvement of the traditional fishers' income. In addition to material and quality of life benefits, these systems have a major role in fostering reinvention of a marine cultural identity for communities. The revived values, norms and cultural symbols (i.e. traditional ceremonies) of sawen and sasi have reinvented the marine cultural identity of Lombok and Maluku people, respectively, and have tangibly restored community pride in their way of life. This implies that fisheries are not considered just as a livelihood, but also as a way of life, a culture and a worldview. As part of that process, local marine ecological knowledge may become integrated in fisheries management. The use of pre-existing fisheries management systems also can stimulate a revival of local traditional ecological knowledge and its use as a complement for common or conventional scientific knowledge.

Nevertheless, major adjustments are required to adapt these pre-existing elements to present-day conditions. Three main aspects need to be examined. The first is that the awig–awig requires an enhanced institutional legitimacy to ensure its wider acceptance. Second, and as was demonstrated by the discourse on sasi in particular, pre-existing systems can mean different things to different people. Therefore a strenuous effort is required to ensure that the various stakeholders accept them as a legitimate, community-based form of resource management. To do that also requires adapting the sasi, for example, to accommodate the biological and ecological parameters that are also essential in comprehensive marine resources management. Third, and extremely challenging, is the need to separate pre-existing systems from some of aspects of their original social context. In Kei Besar Island for example, control over petuanan laut has been the issue of conflict between the 'nobles' and the 'commoners,' two of the three distinct stratifications in Kei society (nobles [*mel*], commoners [*ren*] and slaves [*ri*]). Ownership is a token of the relationship between these classes (Adhuri 1998, 2002b). One likely problem is that because they are embedded within the social construction of the community, under particular contexts petuanan and sasi can be manipulated for social purposes that might be contrary to their functions as an instrument of resources management. As a result, a major adaptation would be required and a strong consensus needed that would function to separate petuanan and sasi from their pre-existing social functions, and to enable them to function in the modern context.

References

Adhuri, D. S. (1998). Saat sebuah desa dibakar menjadi abu: hak ulayat laut dan konflik antar kelompok di Pulau Kei Besar (When a village was burnt to ashes: Communal marine tenure and social conflict in Kei Besar). *Antropologi Indonesia*, 57, 92–109 (in Indonesian).

Adhuri, D. S. (2002a). Selling the sea, fishing for power: A study of conflict over marine tenure in the Kei Islands, Eastern Indonesia. Dissertation, The Australian National University.

Adhuri, D. S. (2002b, June). *From 'old' to 'contemporary' sea wars: analysing conflicts over sea resources, calculating the challenges for fishery co-management*. Paper presented at the EDEN II Workshop: Sustainability and depletion in island Southeast Asia: Forest and fisheries, past and present, Leiden.

Adhuri, D. S. (2005). Menjual laut, mengail kekuasaan: studi mengenai konflik hak ulayat laut di Kepulauan Kei, Maluku Tenggara. (Selling the sea fishing for power: A study on conflict over marine tenure in Kei Islands, Southeastern Maluku). *Masyarakat Indonesia, 21*(1), 127–150.

Anon. (1991). *Laporan penelitian hak adat kelautan di Maluku (Report on marine traditional rights in Maluku)*. Ambon: Yayasan Hualopo, Fak. Hukum dan Pusat Studi Maluku Unpatti (in Indonesian).

Antunès, I. (2000). Le développement local de la pêche en Indonésie, entre unité politique et diversité culturelle. Une approche à partir de deux cas d'étude contrastés, Bendar à Java et Watlar aux Moluques. Dissertation, Université de Paris-IV Sorbonne and University of Sydney.

Antunès, I., & Dwiono, S. A. P. (1998). *Watlar, an Eastern-Indonesian village caught between tradition and modernity*. Monpellier: Centre Orstom.

Bachtiar, I. (2002, Sept). *A strategy of awig–awig development in resources management in Nusatenggara Barat* (translated). Paper presented at the Workshop on Awig–Awig of Fisheries Resources Management, Mataram.

Benda-Beckmann von, F., von Benda-Beckmann, K., & Brouwer, A. (1992, Aug). *Changing 'indigenous environmental law' in the Central Moluccas: Communal regulation and privatization of sasi.* Paper presented at the Congress of the Commission on Folk Law and Legal Pluralism, Wellington.

Kamardi. (1999, July). *Kearifan tradisional dan aspek ekologis* (*Traditional wisdom and ecological aspect*). Paper presented at the Seminar Pemberdayaan Masyarakat Adat Nusa Tenggara Barat, Mataram.

Kissya, E. (1995). *Sasi aman Haru-ukui (Traditional management of sustainable natural resources in Haruku).* Jakarta: Sejati Foundation.

Lokollo. (1988). *Hukum sasi di Maluku: Suatu potret binamulia lingkungan pedesaan yang dicari pemerintah. (Sasi Law in Maluku: A portrait of rural environmental development that is looked for by the government).* Ambon: Faculty of Law, University of Pattimura (in Indonesian).

Lokollo. (1994). *Asas-asas hukum adat kelautan dan manfaatnya bagi pembinaan peraturan daerah di Kabupaten Maluku Tengah dalam rangka implementasi undang-undang nomor 4 tahun 1982 dan undang-undang nomor 9 tahun 1985* (*The foundation of traditional marine law and its functions for a better implementation of Law No. 4/1982 and Law No. 9/1985*). Ambon: Faculty of Law, University of Pattimura (in Indonesian).

Monk, K. A., De Fretes, Y., & Reksodiharjo-Lilley, G. (1997). The ecology of Nusa Tenggara and Maluku. In *The ecology of Indonesia* (Vol. V). Singapore: Eric Oey.

Naamin, N. & Badrudin, M. (1992). The role of coastal village communities and fishermen's organization in the management of coastal fisheries resources in Indonesia. In Anon (Ed.), *Proceedings of FAO/Japan expert consultation on the development of community-based coastal fishery management system for Asia and the Pacific volume 2.* FAO: Rome

Ostrom, E. (1990). *Governing the commons: The evolution of institutions for collective action.* Cambridge: Cambridge University Press.

Pannell, S. (1997). Managing the discourse of resource management: The case of sasi from 'Southeast' Maluku, Indonesia. *Oceania, 67*, 289–307.

Pusdi-PSL Unpatti. (1995). *Kajian hukum tentang norma adat dalam perlindungan lingkungan (Legal study on traditional norms in environmental protection).* Ambon: Kantor Menteri Negara Lingkungan Hidup RI dengan Pusdi-PSL Universitas Pattimura (in Indonsian).

Rahail, J. P. (1995). *Bat Batang Fitroa Fitnangan: tata guna tanah dan laut tradisional Kei. Vol. 4, Seri pustaka khasanah budaya local (Bat Batang Fitroa Fitnangan: Kei traditional land and sea management).* Jakarta: Yayasan Sejati (in Indonsian).

Ruddle, K. (1998). The context of policy design for existing community-based fisheries management systems in the Pacific Islands. *Ocean and Coastal Management, 40*(4), 105–126.

Saad, S. (2003). *Politik hukum perikanan (Fisheries legal politics).* Jakarta: Lembaga Sentra Pemberdayaan Masyarakat (in Indonsian).

Satria, A. (2007a). Sawen: Institution, local knowledge and myths in fisheries management in North Lombok, Indonesia. In N. Haggan, B. Neis, & I. G. Baird (Eds.), *Fishers' knowledge in fisheries science and fisheries management* (pp. 199–220). Paris: UNESCO.

Satria, A. (2007b July). *Do the fishers own their coast? Challenge to devolution of fisheries management: Indonesian perspective.* Paper presented at the Conference on People and the Sea, Amsterdam.

Satria, A., & Matsuda, Y. (2004a). Decentralization of fisheries management in Indonesia. *Marine Policy, 28*, 437–450.

Satria, A., & Matsuda, Y. (2004b). Decentralization policy: An opportunity for strengthening fisheries management system. *The Journal of Environment and Development, 13*(2), 179–196.

Satria, A., Matsuda, Y., & Sano, M. (2006). Questioning community based coral reef management systems: Case study of awig–awig in Gili Indah, Indonesia. *Journal of Environment, Development and Sustainability, 8*, 99–118.

Satria, A., Umbari, A., Fauzi, A., Purbayanto, A., Sutarto, E., Muchsin, I., et al. (2002). *Menuju desentralisasi kelautan*. Cidesindo: Jakarta (in Indonsian).
Wahyono, A., Laksono, D. S., Antarika, I. G. P., Masyhuri, I., Ratna, I., & Sudiyono, S. A. (2000). *Hak ulayat laut di kawasan Timur Indonesia*. Yogyakarta: Media Pressindo (in Indonsian).
Zerner, C. (1991, September). *Imagining the common law in Maluku: Of men, molluscs, and the marine environment*. Paper presented at The Second Annual Meeting of the International Association for the Study of Common Property, Winnipeg.
Zerner, C. (1992, June). *Community management of marine resources in the Maluku Islands*. Paper prepared for FAO/Japan Expert Consultation on the Development of Community-Based Coastal Fishery Management Systems For Asia and the Pacific, Kobe.
Zerner, C. (1994). Through a green lens: the construction of customary environmental law and community in Indonesia's Maluku islands. *Law & Society Review, 28*(5), 1079–1122.

Chapter 3
Open to All?: Reassessing Capture Fisheries Tenure Systems in Southern Laos

Ian G. Baird

Abstract Capture fisheries are among the many different common property resources in the Mekong River Basin described as being historically 'open access'. It is widely accepted that this continues to be the case, and that a 'tragedy of the commons' is therefore inevitable. The myth that all fisheries resources in southern Laos were historically 'open access' is challenged in this chapter. Using the examples of the fence-filter trap and wing-trap fishery system in the Khone Falls area of Khong District, Champasak Province, the operation of fence-filter and wing traps along perennial and seasonal streams in southern Laos, and the pit-trap fishery system in Pathoumphone District, Champasak Province, it is demonstrated that pre-existing tenure systems for fisheries management are far from being 'open access'. Rather, private resource ownership is socially and culturally sanctioned in these ecologically and socially very different fisheries, as part of a common property management system based on first claims to fishing sites and labor inputs. 'Open access' rarely exists in southern Laos, something that may not become evident without sustained field observations. However, discursively characterizing fisheries as 'open access' can be used to justify interventions by government and outside agencies.

Keywords Common property resources • Inland fisheries • Mekong River • Open access • Tragedy of the commons

3.1 Introduction

In the Mekong River Basin many different common property resource management systems, including those for capture fisheries, are typically described as being historically 'open access'. This essentially implies that from pre-modern times until now those resources have been open to exploitation by all, without risk

I.G. Baird (✉)
Department of Geography, University of Wisconsin-Madison, 160 Science Hall, 550 North Park St., Madison, WI, USA 53706-1491
e-mail: ibaird@wisc.edu

K. Ruddle and A. Satria (eds.), *Managing Coastal and Inland Waters: Pre-existing Aquatic Management Systems in Southeast Asia*, DOI 10.1007/978-90-481-9555-8_3,

of being prohibited, limited or sanctioned (Fujita and Phanvilay 2008; Baran et al. 2007a, b; Viner et al. 2006; van Zalinge 2002; ADB 1997). Some believe that many fisheries in Laos are 'open access' (MRC 2006; MRCS 2001; Choulamany 2000; Lorenzen et al. 1998). For most fisheries managers, 'open access' equates with 'unmanaged', implying that resources are therefore extremely vulnerable to overexploitation and depletion, and urgently requiring management interventions from outside. For example, a report produced by the Mekong River Commission Secretariat (2001: 1), as part of an environmental program training case study, stated that, "[o]pen access fisheries are typically characterized by excess harvesting and unsustainable fishing practices."

The idea that local, pre-existing fisheries management measures are unsustainable and thus inadequate, sets the discursive conditions that make outsider-initiated changes in management wholly justifiable, even righteous, by managers and their advisors who see inland fisheries resources as potentially subject to a 'tragedy of the commons' (Hardin 1968). The framework people use for assessing fisheries greatly affects the way they imagine such concepts as 'fish declines' (Bush and Hirsch 2005). Hardin's ideas have profoundly influenced the ways people view fisheries management, approaches to natural resource management in general, and how these subjects are taught.[1] Foucault's (1991) work on 'governmentality' demonstrates how the discursive aspects of constructing fisheries management issues are crucial for understanding how governments and others choose to address particular situations.

In this chapter I argue against the generally dominant assumption that inland fisheries resources in the Mekong Region were historically mostly 'open access', and are thus in need of fundamental restructuring to stop them from being managed as 'common property resources'. Further, I demonstrate that inhabitants of the Mekong River Basin in southern Laos have developed complex systems of fisheries management that are far from being the typically assumed 'open access'.

First I examine some ideas related to common property management and social theory. I then present examples of three important and long-established fisheries that always have been based on restricting access. The first is the fence-filter and wing traps fishery for small migratory cyprinids in the mainstream of the Mekong River, in the Khone Falls area of Khong District, Champasak Province, in southern-most Laos. The second is the fence-filter and wing trap fisheries in seasonal and perennial streams in southern Laos. The third is the use of pit-traps for fishing in the back-swamps of eastern Pathoumphone District, Champasak Province.

---

[1]Hardin argued that common property resources are fundamentally vulnerable to overexploitation because individual users tend to maximize personal benefits, even when the resources they depend on are being overexploited and depleted by the combined actions of users striving to maximize individual benefits, both at the expense of the resource and ultimately of their own interests.

3.1.1 The Tragedy of the Commons

A large international and inter-disciplinary literature that developed during the 40 years since Hardin's (1968) tragedy of the commons thesis appeared has vigorously debated the value of his ideas. There have been many attempts to refute its fundamental premises. Since Hardin was a natural scientist, and people tend to be convinced by arguments presented by those with similar backgrounds, it is not surprising that his thesis has been extremely influential among biologists, geographers, natural resource managers, and the general public. Many have been attracted by the straightforward assumptions in Hardin's theory.

A major problem is that Hardin's thesis tended to justify the alteration of common property management systems to those based on either state control (i.e. such as government managed and enforced protected areas, or government controlled access to resources via quotas or closed exploitation periods) or private control of resources through various means (i.e. community management, private ownership and company control, among others). Essentially, Hardin's work has influenced many to believe that common property systems should be altered to 'non-traditional' forms of management that more effectively 'limit access' to resources (Ostrom 1990; Andelson 1991; Steins et al. 2000).

On the other hand the fundamental and crucial flaws in Hardin's thesis have been noted. First, it essentially collapses 'common property' and 'open access' resources into a single category, reducing many long-standing common property systems to simplistic systems that allow unfettered access to all-comers. Subsequently it has been widely demonstrated that most common property managements systems are far from being 'open access'. Access is often restricted through overt rules and regulations, as well as informal rules and everyday social and cultural norms. This includes practices not formalized in law or mandated by the state. More commonly, however, limitations to access are embedded in socio-cultural practices, including subtle norms (Ostrom 1990; Andelson 1991; Steins et al. 2000). These realities are frequently difficult to recognize without in-depth study of particular circumstances, since important nuances are embedded in local historical, political, economic, cultural, social, and ecological issues.

In addition, a large literature has emerged showing that people are not simply the dominant, individual profit or benefit-maximizing resource exploiters assumed by Hardin. Instead, behavior is almost always mitigated by various social and cultural networks and practices (Bravo and Marelli 2008; Steins et al. 2000; Tanner 2007; Wagner and Talakai 2007). That people everywhere, and not just those living in isolated, inward looking villages of the South, but also urban inhabitants of the industrial economies of the North, are often irrational social animals can be illustrated by tipping behavior in up-scale North American restaurants. Although there are no formal requirements for customers to tip, nevertheless they usually add 10–20% of the meal price to their bill as a tip. If not, it would probably be assumed they were displeased with either the food or the service, or both. Whereas for regular customers this could be construed as rational behavior to ensure future good treatment, it does not explain such behavior by a one-time only patron. In other words, there is no obvious rational economic reason for their leaving a tip. Were people to always act as Hardin's argument assumes it could be reasonably anticipated that all one-time patrons would not leave a tip, thereby maximizing individual benefits. That the opposite behavior is the norm indicates that there exist strong, socially and culturally embedded reasons for tipping (cf. Azar 2003).

3.2 The Fisheries

3.2.1 Fence-Filter (Tone) and Wing Traps (Li) in the Mekong River

Khong District, in Champasak Province, is well known for the Khone Falls, the only large waterfalls on the Lower and Middle Mekong River (Daconto 2001) (Fig. 3.1). Located just north of the Laos–Cambodia border, the Khone Falls area is famous for supporting some of the most important fisheries in the Mekong River Basin. Two of these are the fence-filter trap (tone) and wing trap (li).[2]

3.2.2 Khone Falls Tone Trap Fishery

This fishery targets small cyprinids, known collectively as 'white fishes',[3] of which at least 32 species are believed to annually migrate over 400 km up the Mekong from the Tonle Sap River and Great Lake, in Cambodia. These migrations are linked to lunar cycles, with most fish passing through the Khone Falls to the middle Mekong at the beginning of the new moon. In Laos, fish are expected to arrive during the Chinese New Year, i.e., the beginning of the third month of the Laotian lunar calendar. This period typically falls during a dry season, between late-January and early-February. The fish tend to migrate upstream in the daytime, and the fishery typically lasts for a couple of months, with peak catches during new moon periods, or a little after. Trap catches are low when fish are not migrating. This fishery is among the most important in the Khone Falls area, meeting both subsistence and income needs (Baird 2001; Baird et al. 2003; Baran et al. 2005).

[2]See Claridge et al. (1997) for detailed descriptions of the fishing structure designs for each, as the traps built in the Khone Falls are not identical to those with the same names elsewhere in Laos.

[3]'White fishes' refers to mainly cyprinids that inhabit large rivers and streams. Their colors are often considered to be 'whiter' than 'black fishes' (Channidae, Anabantidae, Claridae and Heteropneustidae, constitute this group) found in more stagnant swamp-like conditions.

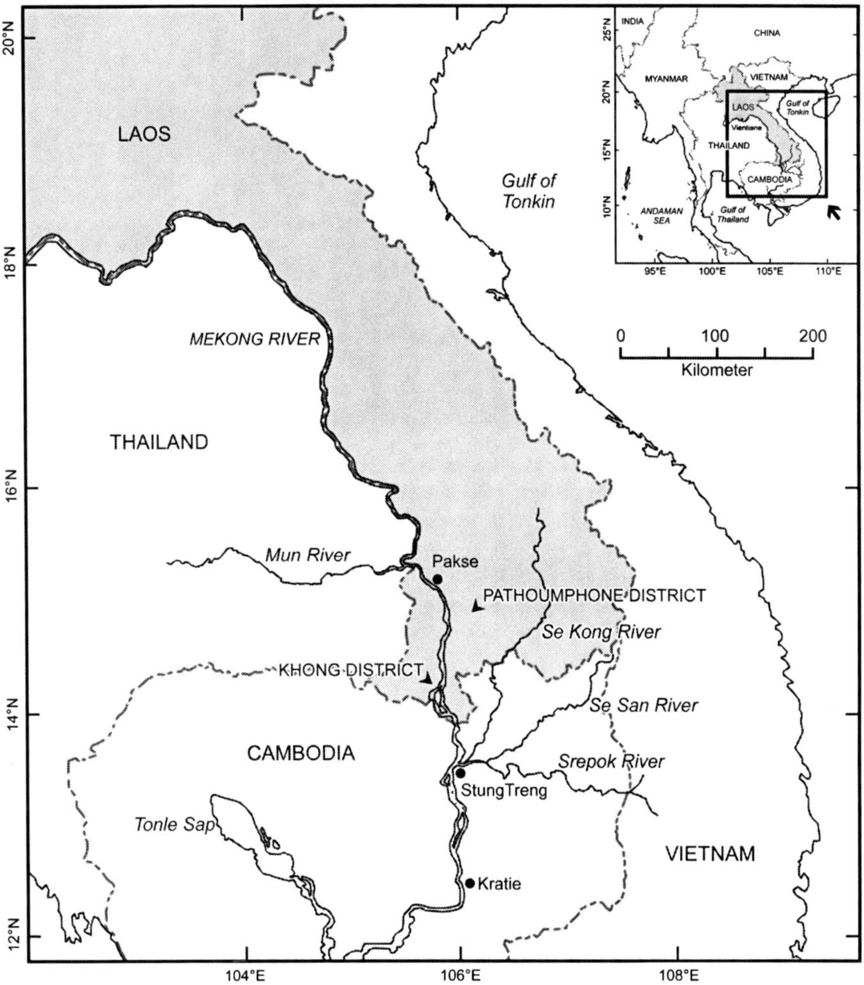

Fig. 3.1 Locations in Laos

The fish tend to move up different channels, and if ascent is blocked by waterfalls or rapids, they return downstream, and are caught in the tone trap. Since these traps catch only fish moving downstream, the Khone Falls is the only place where this kind of fishery can be established, as it only works if rapids or waterfalls are located directly upstream. The main species caught by this fishery are *Henicorhynchus lobatus* (45%), *Paralaubuca typus* (33%), *Henicorhynchus siamensis* (5%), *Labiobarbus leptocheilus* (5%), *Botia modesta* (4%), *Crossocheilus reticulatus* (1%), *Cirrhinus microlepis* (1%), *Tynnichthys thynnoides* (1%), and *Lobocheilos melanotaenia* (1%) (Baird et al. 2003).

The tone fishery involves building trap structures in the rapids of the various channels below the main waterfalls. These structures are typically made of wood,

Photo 3.1 Fence-filter trap (tone) in the Don Nok Kasoum Channel during the dry season to catch migratory small cyprinids, Khone Falls, Khong District, Champasak Province, Laos

bamboo and rattan (Photo 3.1). Various trap designs are used, each based on past observations of the nature of local fish migrations and catches.[4] Many structures are rudimentary, and strong ones are generally not required because water flows are low during the season of this fishery.

3.2.3 Khone Falls Li Fishery

The li fishery in the Khone Falls area primarily targets schools of medium-sized pangasid catfish migrating up the Mekong River, and smaller quantities of other species migrating downstream. However, as with the Khone Falls tone fishery, even if they are trying to migrate upriver, fish are caught when moving downstream to seek alternative routes to avoid rapids or waterfalls. The main species of catfish caught are *Pangasius conchophilus* (41%), *Pangasius krempfi* (5%), *Pangasius bocourti* (4%), *Pangasius larnaudii* (3%), and *Pangasius macronema* (1%) (Baird et al. 2004). The main cyprinids taken are *Henicorhynchus lobatus* (13%), *Scaphognathops bandanensis* (7%), *Cosmocheilus harmandi* (4%), *Cyclocheilichthys enoplos* (3%), and *Puntioplites falcifer* (2%). Over 100 species have been recorded as being caught by this fishery (Baird et al. 2004).

Unlike the dry season tone fishery, these migrations of pangasid catfish and the other species taken are not linked to lunar cycles. Rather, they are associated with hydrological changes and rising waters in the Mekong River. The fishery takes

[4]People in Khong District are famous for their local ecological knowledge (termed 'intimate knowledge' by Raffles (2002) about Mekong fish and fisheries (Baird 2007).

Photo 3.2 Fence-filter trap (tone) in the Xang Pheuak Channel. Khone Falls, Khong District, Champasak Province, Laos

place from May to July, as the waters rise with the onset of the monsoon. After that many traps are either flooded or washed away by the powerful current at the height of the monsoon (Baird et al. 2004; Baran et al. 2005; Hogan et al. 2007).

Like the tone fishery, the li trap structures are made of wood, rattan, bamboo and nails. Li traps are typically much larger and stronger than tone traps, because they must withstand the strong current and higher water levels of the rainy season (Photo 3.2). They are typically built in March and April, when water levels are at their lowest, to be ready for use when water levels rise in May. Building a li trap is hard work that generally takes a group of people many days or even a month or more to complete (Roberts and Baird 1995).

3.2.4 Khone Falls Tone and Li Tenure

The tone and li fish traps in the Khone Falls area are typically owned by ethnic Lao families living in the surrounding villages. Essentially, a tenure right is established by a 'first-comer', with the first person to claim a fishing spot owning in perpetuity the rights to the same fishery at the same time each year (Roberts and Baird 1995; Baird et al. 2003, 2004). In the past, ownership of rice fields was familiar to all, although people had no title deeds. Similarly, fishing spot ownership was known. However, whereas land rights have received formal State recognition, fish trapping spots have not. Nevertheless, government officials in Khong District recognize that people have tenure over certain fishing spots, and the officials have ownership lists. Generally, they uphold these claims, thereby recognizing pre-existing ownership regimes.

The government does not tax the fish caught using tone and li traps, but owners are taxed based on the amount of timber used to build them. Thus, forestry officials visit the traps, assess the amounts of wood used, and charge the users forest use tax. For this reason the Forestry Section of the Khong District government has remained involved in capture fisheries management, despite the task having been allocated nationwide since the 1990s to the Livestock and Fisheries Section. Both Forestry and Livestock and Fisheries are, however, under the district Agriculture and Forestry Office.

For as long as anyone can recollect, tone and li fishing sites in the Khone Falls area have been considered *moun moladok* (inheritable property). Therefore once a claim to a particular fishing site has been established, typically after one season of use, its de facto owner can choose to either use the site privately by himself, team up with others to use it together (in which case catches are divided equally), or rent it to others, either people from the same village or those from other communities. Fishing spots can also be bought and sold like other private property. And, like agricultural land and other private property, they can be divided among successors.

Unlike rice fields, which are generally recognized as being owned year-round, fishing spots are owned only for particular seasons. For example, a tone fishery and a li fishery in the same location might not be owned by the same person(s). However, this is not problematic because the fisheries are seasonally separated. Frequently, trap sites are owned by groups of people or families, sometimes but not always related. Sometimes ownership shares in fisheries are sold or traded to others, while other shareholders retain their shares.

Ownership of tone and li trap sites has also long been important in determining where people could establish new sites. For example, social pressure would be exerted to prevent a person from establishing a new trap site just upstream of an existing one, for fear that it would end up with much of the historic catch of the trap downstream. However, there are limits to the ability for downstream trap owners to prevent those upstream from building new traps. There are some tensions over these matters, which local people appear constantly to be negotiating.

In the Hou Sahong channel of the Khone Falls area there is a special tone fishery targeting the migratory catfish *Pangasius macronema*. Although most traps in the area are privately owned, the village of Hang Sadam communally owns a long-standing fishing site at *Tat Pho*, adjacent to their village. Participation in the fishery is restricted to village members, who work together to make the traps and who receive nearly equal shares of the catch. The system appears to function well, since there has been apparently little dissention among users, or calls for changes in either resource tenure or management (Baird, Hogan et al. 2001).[5]

---

[5]It should be noted that there are plans to build the Don Sahong Hydroelectric dam on the Hou Sahong channel in the general area where the communal fishery for *Pangasius macronema* is located (see International Rivers 2008).

3.2.5 *Fence-Filter Trap (Tone) and Wing-Trap (Li) Fishing Along Seasonal and Perennial Streams*

Unlike the fence-filter and wing trap fisheries specifically designed for a particular habitat on the mainstream of the Mekong River in the Khone Falls area, the fisheries described in this section occur in various forms, designs and sizes, and in a wide variety of different locations throughout Laos. They are found almost everywhere streams are used by migratory fish at the end of the rainy season.[6]

3.2.5.1 Stream Tone and Li Fisheries

Unlike the highly specialized fisheries at the Khone Falls, many other less-spectacular kinds of fence-filter and wing traps are common and widespread through Laos. They are typically important for rural livelihoods (Photo 3.3).

There are some fundamental differences between the Khone Falls fence-filter trap and wing trap fisheries and those elsewhere. For example, whereas the Khone Falls traps are active in the dry season, early in the calendar year, stream fence-filter and wing traps are operational during the rainy season. Although some people use these traps at the beginning of the rainy season to catch fish migrating upstream into natural ponds and wetlands to spawn, they are used especially at the end of the rainy season. Another major difference between the fence-filter and wing traps in the Khone Falls area and those in streams is that whereas the former target

Photo 3.3 Wing trap (li) during the rainy season to catch mainly migratory catfish, Khone Falls, Khong District, Champasak Province, Laos

[6]See Claridge et al. (1997) for detailed descriptions of these types of traps.

long-distance migratory 'white fish', stream traps target 'black fish' migrating short distances from seasonal to perennial water bodies. These include *Channa striata*, *Clarias* spp., *Trichogaster trichopterus*, *Puntius brevis*, *Hampala dispar*, *Anabas testudineus*, *Ompok bimaculatus*, *Rasbora* spp., and others. Thus the species composition of both fisheries is almost totally different. Another crucial difference is that the fish caught in the Khone Falls are migrating upriver when trapped, whereas those caught at the end of the rainy season are all moving downstream to exit the progressively drying wetlands and streams.

3.2.5.2 Stream Tone and Li Tenure

Although there are significant differences in the two types of fence-filter trap and wing trap fisheries discussed here, the tenure arrangements associated with stream tone and li traps are quite similar to those involving tone and li traps in the Khone Falls. That is, stream tone and li traps are privately owned, and ownership is based on who fishes first in an area using a fixed trap.[7] Once tenure is established, others cannot erect similar types of traps during the same fishing season in the direct vicinity of the already established trap site. They may be able to do so if their trap is located a significant distance from the other.[8] Thus, the trap site comes to be regarded as private property that can be transferred like land, and which is inheritable, just like fence-filter traps and wing traps used in the Khone Falls area.

However, in some cases fence-filter trap and wing trap tenure is affected by land tenure issues, which is not generally the case in the Khone Falls. If a trap site is located in a commons area, adjacent to common forest areas for example, the above formula for establishing tenure would almost certainly be applied. However, access is further limited if a fishing site is located adjacent to someone's private land. In such cases people may argue that their right to operate fish traps in the stream is dictated by their ownership over adjacent land. Thus land rights and fishing trap rights can sometimes be linked. The claim can be especially strong if a farmer owns the land on both sides of the stream in question. In those cases it would be hard for someone to fish at a particular location without first obtaining permission from the owner of the adjacent land. However, if fish trap ownership preceded land ownership, then the rights of the land owner would generally be greatly reduced. Of course, the unique ecological, social, cultural, political and economic circumstances all greatly affect the forms that tenure arrangements take.

The management of stream fish traps is generally done entirely by local people, with very little interference from state agencies. This sort of fishing is considered to be 'traditional', and therefore local people see the use of these traps as an inherent right, just like growing rice. People often use them to catch the raw materials for

---

[7]Mobile gear often involves different rules and norms.

[8]This distance can vary depending on various ecological, geographical and social factors.

making fermented fish paste (*pa dek*), a culinary staple. In addition, these traps tend to catch smaller amounts of fish compared to fence-filter and wing traps used in the Khone Falls. Stream trap catches can, however, range from just a few kilograms of small fish to hundreds of kilograms of fish, including larger, more expensive table species. The government is not interested in generating revenue from these smaller fishing operations, thus they are rarely if ever taxed. The only exception might be if a particular trap was well-known for catching particularly large quantities of fish. Then, the state might demand some of the benefits. The community might also request a cut. There are various possibilities in a diverse region such as this.

3.2.6 Pit-Trap (loum pa) Fishing in Swamps

The *'Khet Beung'* area of eastern Pathoumphone District, Champasak Province, is located tens of kilometers east of the mainstream Mekong River, and away from any other large rivers. Unlike the Khone Falls area, the *'Khet Beung'* area is not known for its many islands and multiple channels, and it is not the type of average place where stream traps are used. It is a special place with rich forests and plentiful natural wetlands, the largest of which are known as *'beung'*, Eastern Pathoumphone is known as a 'wetlands region' and is one of the largest wetland complexes in Laos (Claridge 1996). Seventy percent of the district's land is included in two of Laos' National Protected Areas (NPAs), the Xepian NPA in the southeastern part of the district, and Dong Houa Sao NPA, in the northeastern part of Pathoumphone.

The ecological and socio-cultural circumstances of the Khet Beung area differ considerably from those in the Khone Falls area. For one, the long distance migratory fish do not occur, whereas 'black fish' similar to those caught in stream traps are common. On average, the most abundant species in pit-traps, based on percentage of total weight, are *Channa striatus*, followed by *Monopterus albus*, *Clarias batrachus*, *Trichogaster trichopterus*, *Puntius brevis*, *Hampala dispar*, *Anabas testudineus*, *Ompok bimaculatus*, and *Rasbora* spp. Snakes, crabs and snails are also harvested in smaller quantities. None of the main species caught in this fishery are the same as dominant species in the Khone Falls area, although many are the same as those caught by stream traps.

According to surveys done in 2006 (the results of which are unpublished), each pit-trap yields 15–30 kg of fish. Some yield less fish and more *Monopterus albus*, others yield more *Channa striatus*, whereas in others *Clarias batrachus* are the most common, depending on the particular ecological conditions involved. Often, predatory fish have eaten many of the smaller species before the areas are harvested, but in some cases early harvests can result in higher catches of small minnows and carps.

Unlike at the Khone Falls, the ethnic Lao people and minorities who live in Pathoumphone do not wait for the Chinese New Year for fish to migrate up the Mekong River. Neither do they catch their fish at the end of the rainy season, like those using stream traps. Instead, they anticipate declines in water levels as the dry season arrives. They wait for the many perennial and seasonal natural ponds that dominate the landscape of this region to partially or fully dry out, then they take advantage of the ecological conditions to catch fish. Harvesting tends to occur at the height of the dry season.

The most important way that villagers take advantage of the ecological conditions of the Khet Beung area to catch large amounts of fish is through the adoption of a long-standing fishing method particular to the area and its ecological conditions: pit-trap fishing (*loum pa*, in Lao). This involves digging deep pits (often one or more meters deep and a meter or more in width and about the same in length) at the bottom of perennial or seasonally inundated wetlands during the dry season. The pits are then filled with sticks and other vegetation before they are inundated during the rainy season. When water levels begin to drop naturally at the end of the rainy season, in October or November, fish move to deeper waters, including the pit-traps that villagers have set up. Depending on the individual ecological conditions of each wetland and pit-trap, once water levels have declined considerably, especially from December to February, villagers remove the sticks and other debris from the pit-traps, and scoop out the water so that the fish can be caught easily.[9]

3.2.6.1 Pit-Trap Tenure

As with the tone and li fisheries, the pit-trap fishery in Pathoumphone is far from being 'open access'. Each pit-trap is considered to be privately owned by villagers living either in the village closest to the area, or in neighboring villages. Tenure over the pit-traps is obtained through labor inputs. Once a pit-trap has been created, it is owned in perpetuity by the digger or the digger's family. As with tone and li fishing spots, pit-traps are considered to be inheritable property, and they too can be rented out and sold like other private property.

Pit-trap tenure can be compared with that applied to wood resin trees (*Dipterocarpus alatus*) throughout much of Southeast Asia. In most places ownership of a wood resin tree is based on who first tapped it (Baird and Dearden 2003; Baird 2009). The same principle applies in Pathoumphone, where a pit trap is owned by the person who first dug it. Similarly, Baird and Bounphasy (2005) found that wild honey bees in Pathoumphone are managed privately, based on ownership of their nests. However, unlike pit-traps and wood resin trees, private ownership is not permanent. People who discover bee nests in the forest, mark the trees to claim ownership of the nests during that dry season. Others may not exploit bee nests in a tree already marked. Then, at the end of the season, when all the nests have been harvested, the marked trees revert to being the common property of the village.

---

[9]More recently, there have also been limited reports of people using gasoline powered pumps to remove water from pit-traps.

There are apparently no restrictions on how many pit-traps a person, family or household can own in Pathoumphone, and villagers typically report that the number depends largely on the labor available for harvesting fish and maintaining the traps. Frequently, families own between a few and over ten pit-traps. It might be expected that numbers would be restricted, because the catch of each is probably affected by the total number in the same wetland. However, informants report that the number of private pit-traps is not restricted. Essentially, common wetlands can be partially privatized by digging pit-traps. However, their owners have no tenurial rights over the communal wetland in general as a result; fishing rights are restricted to particular locations at certain times of the year. Therefore tenure rights are strong, but seasonal and partial. Owners also must periodically dredge their traps and otherwise reinforce them with wood, both to maintain fish production and unequivocally establish their tenure over particular pit-traps.

3.3 The Nature of Tenure and Governance

The cases described in the preceding sections represent three examples of clear-cut, long-standing tenure arrangements for fisheries in southern Laos. Despite the very different circumstances associated with each fishery, all three involve complex common property resource management rules and norms that include private tenure arrangements for managing resources. Further, scarcity of good fishing spots characterizes all of these fisheries. The scarce nature of fishing sites has apparently encouraged villagers to develop private tenure systems in all three instances, thereby avoiding conflicts over the basis for production.

However, these cases also raise many important questions about tenure arrangements associated with other fisheries, which may or may not be organized in the same way as tone and li or pit-trap fisheries. They also raise questions regarding restrictions on the use of mobile gears and whether, for example, the use of casting nets and gillnets is allowed in rivers. Although at first glance many net and hook based fisheries appear to be 'open access', which fishers will likely affirm to be the case, the reality might be quite different.

'Open access' implies that outsiders can fish as they wish, and theoretically this should be possible if the system is truly 'open access'. However, few situations in Laos can be accurately described as fully 'open access', although many could be considered partially 'open access'. The term open access is frequently confusedly used to describe situations where common property management systems are in place (Ostrom 1990).

There are also other local fisheries management systems in Laos that regulate access to particular resources. For example, there are many different ways that individuals and communities manage seasonal and perennial wetland fisheries (see Tubtim and Hirsch 2005; Mollot et al. 2007). Mollot et al. (2005) have documented complex arrangements for limiting access to a *Macrobrachium* shrimp fishery in the Nam Khan River of northern Laos. Some important places for setting gillnets in the Khone Falls area are also privately owned, like tone and li traps. The same applies to sites for other types of traps in the Khone Falls (Roberts and Baird 1995).[10]

There is also the matter of protected areas for fish, or fish sanctuaries, which are established in particular areas and subject to varying kinds of restriction, either as a result of long-standing spiritual beliefs, from natural resource management concerns, or as a result of other more recently introduced fisheries management arrangements. In Laos, protected areas for fish have existed for as long as anyone can remember, and have unwritten local rules and norms attached that either ban fishing year round, disallow fishing at certain times of the year, or prohibit particular types of fishing (Baird 2006). Resources cannot be considered 'open access' if either formally or informally protected areas are located within them.

Let us consider what 'open access' commonly really means in the Laotian context, and the discourses associated with the concept. In my experience, usually fishing is open to outsiders, but is almost always restricted. Typically, outsiders are expected to follow particular local rules and norms, which if not respected lead to various forms of protest, like the enforcement of customary rights or use of social pressures. Thus it is apparent even at this level that access is not 'open'; outsiders may be allowed to fish in a village's fishing grounds, but they should do so in ways similar to the villagers. If locals use small-meshed gillnets, outsiders would probably not be prohibited from doing so. However, were outsiders to try using gears that locals never use, or substantially larger or more efficient versions of the gears in use, then the villagers would likely no longer consider their fishing areas to be 'open access'. For example, if the smallest mesh size for gillnets used in a village is 2.5 cm, it is unlikely that they would countenance outsiders using a 2-cm mesh size.

In other words, access is open only if outsiders follow particular norms. However, the discourse of villagers may not always make that clear from the outset, hence the misunderstanding of the situation by outsiders, and therefore the crucial importance of detailed and careful field examination of each fishery.

That villagers frequently describe their fisheries as being 'open access', when they really are not stems from the relationship between rural people and officialdom. Typically, full 'open access' is constrained by local rules or norms that the state regards as informal. Such rules and norms are neither promoted nor sanctioned by government. Rather, they are community approved, which can be important, even if the rules are not legally binding. However, when the state is not explicitly involved, local people commonly describe their resources as 'open access', even when they are governed through complex common property regimes. Essentially, villagers use the term 'open access' to mean that the state

[10]However, this sort of system is not in place for most locations where fishing in the Khone Falls is possible. Only particularly important fishing spots have become privately owned.

does not restrict people, even if the villagers might do so. Considering that at present there is no fisheries law in Laos,[11] and that the state generally leaves local fisheries management to fishers,[12] local rules and norms remain fundamentally important.

Another important aspect of tenure relates to migratory species. Although local resources might not be openly accessed by everyone, Mollot et al. (2007: 61) argue that migratory fish are essentially 'open access' because they move among various independently managed areas:

> In riverine fisheries, the approaches to community-based capture fisheries management may also support access rights of a host village to the exclusion of other neighboring villages, but the fish stocks themselves may be moving between villages across large geographical areas as part of seasonal fish migration cycles. This essentially creates an open access fishery in the rivers that support some of the most important and productive fisheries in the Mekong Basin.

However, it can be argued that even such fish stocks are not 'open access' because access is still restricted in the individual management areas. Two different issues are involved, so using the same term to describe both is confusing and misleading. Fish mobility undoubtedly complicates management. However, that is not related to a resource being 'open access'; it is about a resource being mobile and passing through different management areas, thereby creating management problems that cannot be solved by individual management areas alone.

That fisheries are frequently characterized as 'open access' is related to state and institutional power. Important here is Foucault's concept of 'governmentality', which helps explain why, despite overwhelming evidence to the contrary, bureaucrats and their allies commonly define complex tenure arrangements as 'open access' situations. Most importantly, state discourses are often simplifications of reality that help make management systems more visible and thus controllable by the state (Scott 1998). However, specifically related to Foucault's ideas, state and aid agency intervention in local resource management is often justified by portraying already existing management systems as 'open access', and thus fundamentally flawed and in urgent need of state intervention. In other words, the discourses promoted by bureaucrats tend to make state interventions, whether useful or not, seem like the responsible thing to do, rather than an unwarranted interference in local affairs. Outsider intervention is justified to prevent the tragedy of the commons. As Foucault (1991) points out, it is crucial to consider how state discourses are created and deployed, and how they contribute to maintaining state power and influence. The use of the concept of 'open access' provides a prime example of how the

[11]Although attempts are being made to pass fisheries legislation in Laos, those involved hope to be able to support legislation that facilitates and supports local management efforts, rather than restricting local people from improving their management of fisheries resources. The legislation is expected to be passed in 2010.

[12]This differs from Cambodia, where fisheries are generally more valuable, and where the state is reliant on revenues from them.

deployment of a particular discursive framework can justify the use and misuse of state or other institutional power.

3.4 Conclusions

Some observers have described fisheries in Laos as being historically ‘open access’. However, reality has long been more complex, with elements of private ownership frequently being introduced when deemed useful.

In this chapter I have described three important fishery tenure systems from southern Laos, involving fence-filter and wing traps at the Khone Falls and in streams, and pit-trap fishing in the Khet Beung area, and have illustrated how they have long been managed as common properties. None is a product of efforts by outsiders to improve fisheries management, although historically they would have been influenced by the state in various ways. Crucially, private ownership has long been recognized by ‘first claims’ to good fishing spots, as well as through fishers’ labor inputs to ensure successful fishing.

In addition, these examples of complex tenure arrangements demonstrate that even common fishing activities involving the use of nets and hook-and-line are not ‘open access’, as typically portrayed both by outside observers and the fishers themselves. I have proposed that villagers often miscommunicate information about these tenure arrangements because they distinguish state-sanctioned rules from locally made unwritten rules and norms, with the latter not being considered in some kinds of public discourse. I have also demonstrated that state powers and other outsiders sometimes have a vested interest in falsely characterizing common property systems as ‘open access’, thus enabling them to justify external intervention to forestall what they perceive as an otherwise inevitable ‘tragedy of the commons’.

I do not attempt to argue that all local people manage aquatic resources sustainably, since the various changes associated with human population increase, changes in fishing and post-harvest technologies, and the increased importance of markets have had an impact on the ways people use and manage fisheries resources. Also, as pointed out by Mollot et al. (2007), these changes are compounded because many fish in the Mekong River Basin are highly migratory, thus making it difficult for individual fishing communities to determine how particular fish stocks are or should be managed (see, for example, Baird and Flaherty 2004; Hogan et al. 2007; Baird et al. 2003). Regardless of intent, limited information about other places where fish spend parts of their life cycles makes it difficult for individual communities to determine the future condition of their fisheries, and therefore the strategies for basing management decisions.

In addition, many other serious threats to their fisheries are beyond the control of fishing communities. These include industrial pollution, domestic wastes, land use on critical floodplains, and especially the construction and operation of large hydropower dams in both the Mekong River and its tributaries. However, a balanced perspective is important, and it should be understood that fishing communities can improve their fisheries by implementing local rules, including those for the establishment and management of fish sanctuaries, as has been widely demonstrated (see Baird et al. 2001b; Baird and Flaherty 2005; Baird 2006).

Many serious challenges confront the Mekong River Basin and its inhabitants. On the one hand the region has been blessed with some of the world's most biodiverse and productive freshwater fisheries. On the other, the heavy dependence on these capture fisheries for food and livelihood, especially by the poor, also reveals a potential danger for present and future generations if they are not managed sustainably. Collapse of these fisheries could leave many people without an important safety net, and a resource that people turn to when agriculture fails in the absence of alternative livelihoods. Without a doubt a more sustained vision is needed to ensure that fisheries in the Mekong River Basin are better managed in the future than they are at present.

Finally, I have stressed the importance of considering different scales of management. It is not enough to encourage people simply to improve their management of local fisheries, although that is an important part of what is required. Nor is it enough to analyze just regional concerns, although ignoring these issues would also be foolish. The point of this chapter is that although the future of capture fisheries in Laos and the Mekong River Basin are in jeopardy, inaccurately labeling fisheries there as being simply 'open access' will not make various complex issues easier to address. In fact, such labeling would suggest that there has been no serious effort to understand fully the nature of each fishery. This needs to be rectified quickly, because only when one's gaze is focused and sustained, and power relationships are understood, does the full nature of what one is looking at become evident.

Acknowledgements I thank Somphong Bounphasy and Kaikeo Oudomxay from the Global Association for People and the Environment (GAPE) for adding to my knowledge about pit-trap fisheries operation and management in Pathoumphone District, Champasak Province. I am also grateful for comments provided on an earlier draft by Roger Mollot and Kenneth Ruddle. Responsibility for any remaining shortcomings is mine alone.

References

ADB (Asian Development Bank). (1997). *Asian Development Bank policy on fisheries*. Manila: ADB.

Andelson, R. V. (1991). *Commons without tragedy*. Savage, MD/London: Barnes and Noble Books/Shepheard-Walwyn Ltd.

Azar, O. H. (2003). The social norm of tipping: a review. http://129.3.20.41/eps/othr/papers/0309/0309006.pdf. Accessed 18 Dec 2009.

Baird, I. G. (2001). Aquatic biodiversity in the Siphandone wetlands. In G. Daconto (Ed.), *Siphandone Wetlands. Environmental protection and community development in Siphandone Wetland Project* (pp. 61–74). Bergamo: CESVI.

Baird, I. G. (2006). Strength in diversity: Fish sanctuaries and deep-water pools in Laos. *Fisheries Management and Ecology, 13*(1), 1–8.

Baird, I. G. (2007). Local ecological knowledge and small-scale freshwater fisheries management in the Mekong River in southern Laos. In N. Haggan, B. Neis, & I. G. Baird (Eds.), *Fishers' knowledge in fisheries science and management* (pp. 246–266). Paris: UNESCO.
Baird, I. G. (2009). *Dipterocarpus wood resin tenure, management and trade: practices of the Brao in Northeast Cambodia*. Saarbrücken: Verlag Dr. Müller.
Baird, I. G., & Bounphasy, S. (2005). Non-timber forest product management in southern Laos: The necessity of considering complex tenure arrangements. In Anon (Ed.), *Resource book on enhancing the poor's access to land and other common property resources*. Quezon City: Asian NGO Coalition for Agrarian Reform and Rural Development and the International Land Coalition. Paper 22.
Baird, I. G., & Dearden, P. (2003). Biodiversity conservation and resource tenure regimes – A case study from northeast Cambodia. *Environmental Management, 32*(5), 541–550.
Baird, I. G., & Flaherty, M. S. (2004). Beyond national borders: Important Mekong River medium sized migratory carps (Cyprinidae) and fisheries in Laos and Cambodia. *Asian Fisheries Science, 17*(3–4), 279–298.
Baird, I. G., & Flaherty, M. S. (2005). Mekong River fish conservation zones in southern Laos: Assessing effectiveness using local ecological knowledge. *Environmental Management, 36*(3), 439–454.
Baird, I. G., Flaherty, M. S., & Phylavanh, B. (2003). Rhythms of the river: Lunar phases and migrations of small carps (Cyprinidae) in the Mekong River. *Natural History Bulletin of the Siam Society, 51*(1), 5–36.
Baird, I. G., Flaherty, M. S., & Phylavanh, B. (2004). Mekong River Pangasiidae catfish migrations and the Khone Falls wing trap fishery in southern Laos. *Natural History Bulletin of the Siam Society, 52*(1), 81–109.
Baird, I. G., Hogan, Z., Phylavanh, B., & Moyle, P. (2001). A communal fishery for the migratory catfish *Pangasius macronema* in the Mekong River. *Asian Fisheries Science, 14*, 25–41.
Baird, I. G., Phylavanh, B., Vongsenesouk, B., & Xaiyamanivong, K. (2001). The ecology and conservation of the Smallscale Croaker *Boesemania microlepis* (Bleeker 1858–59) in the mainstream Mekong River, southern Laos. *Natural History Bulletin of the Siam Society, 49*, 161–176.
Baran, E., Baird, I. G., & Cans, G. (2005). *Fisheries bioecology in the Khone Falls area (Mekong River, Southern Laos)*. Penang: WorldFish Center.
Baran, E., Jantunen, T., & Chong, C. K. (2007). *Values of inland fisheries in the Mekong River Basin*. Phnom Penh: The WorldFish Center.
Baran, E., Starr, P., & Kura, Y. (2007a). *Influence of built structures on Tonle Sap fisheries*. Phnom Penh/Penang: Cambodia National Mekong Committee/The WorldFish Center.
Bravo, G., & Marelli, B. (2008). Irrigation systems as common-pool resources: Examples from Northern Italy. *Revue de Geographie Alpine, 96*(3), 5–25.
Bush, S. R., & Hirsch, P. (2005). Framing fishery decline. *Aquatic Resources, Culture and Development, 1*(2), 79–90.
Choulamany, X. (2000). *Fisheries survey, Luangprabang Province, Lao PDR*. LARREC Research Report No. 0001, Vientiane: Living Aquatic Resources Research Centre.
Claridge, G. (compiler) (1996). *An inventory of wetlands of the Lao PDR*. Vientiane: Wetlands Programme, IUCN – The World Conservation Union.
Claridge, G. F., Sorangkhoun, T., & Baird, I. G. (1997). *Community fisheries in Lao PDR: A survey of techniques and issues*. Vientiane: IUCN – The World Conservation Union.
Daconto, G. (Ed.). (2001). *Siphandone Wetlands. Environmental projection and community development in the Siphandone Wetland Project*. Bergamo: CESVI.
Foucault, M. (1991). Governmentality. In G. Burchell, C. Gordon, & P. Miller (Eds.), *The Foucault effect: Studies in governmentality* (pp. 87–104). Chicago: University of Chicago Press.
Fujita, Y., & Phanvilay, K. (2008). Land and forest allocation in Lao People's Democratic Republic: Comparison of case studies from community-based natural resource management research. *Society and Natural Resources, 21*(2), 120–133.
Hardin, G. (1968). The tragedy of the commons. *Science, 162*, 1243–1248.

Hogan, Z., Baird, I. G., Radtke, R., & Vander Zanden, J. (2007). Long distance migration and marine habitation in the Asian catfish, *Pangasius krempfi*. *Journal of Fish Biology, 71*, 818–832.
International Rivers. (2008). *Power surge: The impacts of rapid dam development in Laos*. Berkeley, CA: International Rivers.
Lorenzen, K., Garaway, C. J., Chamsingh, B., & Warren, T. J. (1998). Effects of access restrictions and stocking on small water body fisheries in Laos. *Journal of Fish Biology, 53*(Supplement A), 345–357.
Mollot, R., Chamsingh, B., & Noraseng, P. (2007). From traditional management to fisheries co-management – A review of benefit sharing in community-based fisheries management in Lao PDR. In Anon (Ed.), *A fair share? Experiences in benefit sharing from community-managed resources in Asia* (pp. 51–66). Bangkok and Hanoi: RECOFTC, WWF and SNV Netherlands Development Agency.
Mollot, R., Xainyavi, B. & Sysombath, D. (2005). Local ecological knowledge and customary resource tenure for a *Macrobrachium* fishery on the Nam Khan. In Anon. (Eds.), *Proceedings of 7th Technical Symposium on Mekong Fisheries Ubon Ratchathani, Thailand, 15–17 November 2000* (pp. 321–324). Phnom Penh: Mekong River Commission.
Ostrom, E. (1990). *Governing the commons: The evolution of institutions for collective action*. Cambridge: Cambridge University Press.
MRCS (Mekong River Commission Secretariat). (2001). *Sustainable management of the Nam Ngum Reservoir fishery*. MRCS Environment Training Program Case Studies. Phnom Penh: MRCS.
MRC (Mekong River Commission). (2006). *Livelihoods approaches and fisheries management in the Lower Mekong Basin*. Mekong Fisheries Management Recommendation #5. Vientiane: The Technical Advisory Body for Fisheries (TAB), Mekong River Commission.
Raffles, H. (2002). Intimate knowledge. *International Journal of Social Sciences, 173*, 325–335.
Roberts, T. R., & Baird, I. G. (1995). Traditional fisheries and fish ecology on the Mekong River at Khone Waterfalls in Southern Laos. *Natural History Bulletin of the Siam Society, 43*, 219–262.
Scott, J. C. (1998). *Seeing like a State. How certain schemes to improve the human condition have failed*. New Haven: Yale University Press.
Steins, N. A., Edwards, V. M., & Roling, N. (2000). Re-designed principles for CPR Theory. *The Common Property Resource Digest, 53*, 1–5.
Tanner, A. (2007). On understanding too quickly: Colonial and postcolonial misrepresentation of indigenous Fijian land tenure. *Human Organization, 66*(1), 69–77.
Tubtim, N., & Hirsch, P. (2005). Common property as enclosure: A case study of a backswamp in southern Laos. *Society and Natural Resources, 18*, 41–60.
Van Zalinge, N. P. (2002). Update on the status of the Cambodian inland capture fisheries sector with special reference to the Tonle Sap Great Lake. *Mekong Fish Catch and Culture, 8*(2), 1–9.
Viner, K., Ahmed, M., Bjørndal, T., & Lorenzen, K. (2006). *Development of fisheries co-management in Cambodia: A case study and its implications*. Discussion Series, No. 2. Penang: WorldFish Center.
Wagner, J., & Talakai, M. (2007). Customs, commons, property, and ecology: Case studies from Oceania. *Human Organization, 66*(1), 1–10.

Chapter 4
Seasonal Ritual and the Regulation of Fishing in Batanes Province, Philippines

Maria F. Mangahas

Abstract The *mataw* fishers of Batanes, the ten small northernmost islands of the Philippine archipelago, engage in the seasonal capture of Flying fish (Exocoetidae) and Dorado (*Coryphaena hippurus*), known locally as the 'fish of summer' (*among nu rayon*), that enter the coastal waters in the summer months of March through May. Each fisher is identified by the '*vanua*' or 'port' to which he belongs. The vanua is a specific spatial location, but also a particular organized group led by the fisher chosen to make the 'first fishing trip', to perform ritual and implement the rules of the group. In addition, the ritual schedule governs the use of other gears, thereby regulating fishing activities on traditional grounds. This has potential implications for the stocks of both migratory and demersal species, in terms of closed season, fishing quotas, protected areas, and control over gear use. Via the performance of seasonal rites, which also organize the fishers into a cooperative association with their 'clean vanua', the fish are coaxed to fulfill the fishers' subsistence needs and the ancestral spirits (*añitu*) are called on to bring luck and forestall tragedy. Tension and creative negotiation exists between the values and practices of the *vanuas* inherited from the ancestors and the modern values and ideas that accompany newly introduced technologies and the market.

Keywords Batanes • Fishing rituals • Ritual technology • Marine resource management

4.1 Introduction

Batanes, the smallest province of the Philippines, is made up of ten small islands. Only three, Itbayat, Sabtang and Batan, are inhabited. Batanes Province is located north of the main island of Luzon, from which it is separated by the Balintang

M.F. Mangahas (✉)
Department of Anthropology, University of the Philippines, Diliman, Quezon City 1101, Philippines
e-mail: maria.mangahas@up.edu.ph

K. Ruddle and A. Satria (eds.), *Managing Coastal and Inland Waters: Pre-existing Aquatic Management Systems in Southeast Asia*, DOI 10.1007/978-90-481-9555-8_4,

Channel (Fig. 4.1). The population numbers around 18,000, of whom approximately 65% live on Batan Island, where Basco, the provincial capital, is located. There is little population increase, owing to a high rate of out-migration of young Ivatans (the people of Batanes), who leave for work or study, and then often settle permanently elsewhere in the Philippines.

Batanes is isolated by the dangerous waters of the Balintang Channel. They have protected it from aggressive fishers from elsewhere in the Philippines, so the province has been less vulnerable to the destructive fishing and resource depletion by outsiders that have so damaged other fisheries throughout the country (Aprieto 1995). Further, the fish catch satisfies only local demand, and has not entered significantly into external markets. Local consumption of fresh fish has expanded since the introduction of electricity, in the late-1990s, which enabled many households to own refrigerators. The fisheries of Batanes are now under pressure from local technological innovation

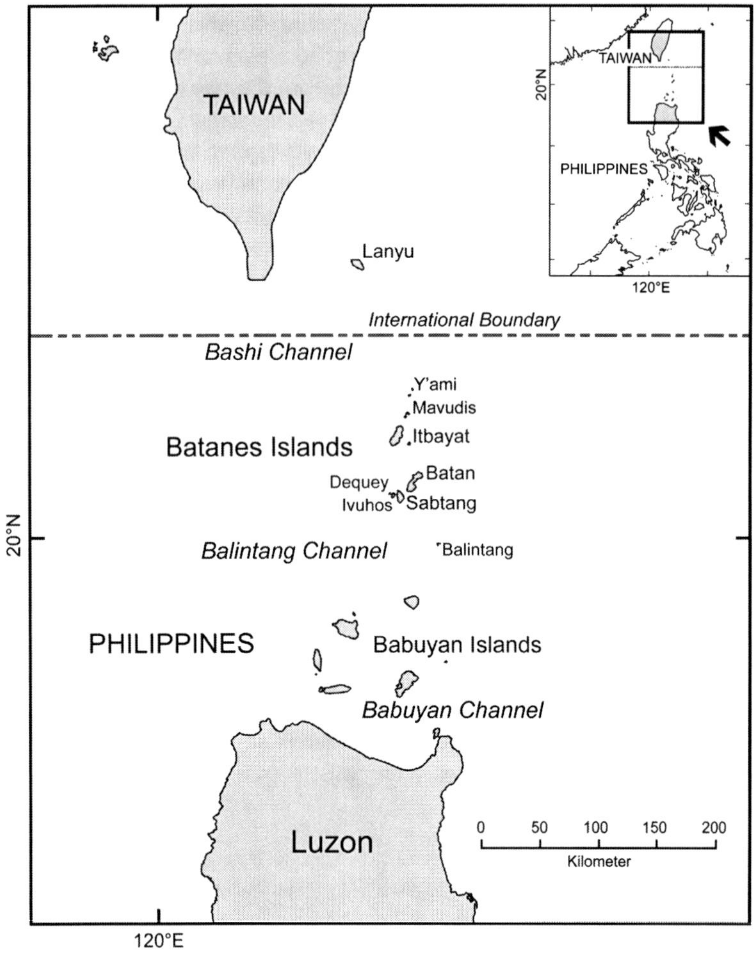

Fig. 4.1 Location of Batanes Province, Philippines

and modernization, including mechanization, drift- and trawl net technologies (Hornedo 2000), and fishing based on compressor-assisted diving.

The Ivatans share mainstream Filipino culture. Christianity was introduced by Dominican missionaries, the first of whom arrived in 1686. *Reduccion*, or the forced reorganization of settlements from scattered hilltops to the lowlands 'under the church bells', and planning of towns around the plaza, church and town hall was implemented in Batanes under Spanish colonial rule.[1] Despite geographical isolation, the difficulties of transport and the rural context of daily life, Batanes now has all the amenities of modern life (Tarrobago 2003).

Ivatans are typically root crop farmers, with the white yam or *uvi* (*Dioscorea alata*), a plant of cultural significance (Datar 1999) grown for both subsistence and pig feed. Households also obtain cash by cultivating such crops as garlic, and by raising cattle. The main sources of wage employment are government offices and agencies, stores and private enterprise, mostly in Basco. Income is derived from tourism, mainly in summer.

Fishing is important in Batanes, particularly in summer, when the weather is generally fair and the seas calm, and also because the seasonal migratory fish, Flying fish and Dorado, enter nearshore waters. Farmer-fishers intent on seizing this seasonal opportunity then set aside or delegate farm tasks, to focus on fishing. Mataw fishers (those who catch Dorado) may even move their families for the duration of the fishing season to temporary houses in their fields that are closer to their 'port', or vanua.

In Batanes, mataw fishing or angling for Dorado is typically done by a lone fisher from a small boat, known locally as a *tataya*, that is rowed and powered by a small sail (Photo 4.1). Mataw fishing involves first catching the Flying fish used

Photo 4.1 A mataw fishing boat coming ashore, Batanes Province, Philippines

[1]The Spanish administration lasted in Batanes for two centuries, from the arrival of Dominican missionaries, in 1686, to the Philippine Revolution, in 1898.

as live bait for Dorado. Scoop net fishing (*sumuho*) with a light for Flying fish involves at least three crewmembers and a slightly larger boat.[2] Using similar methods, fishing in summer for Flying fish and Dorado is also done off Sabtang and Itbayat islands, and in Lanyu (Hsu 1982). However, off Sabtang it is usually done from mechanized and larger boats than off Batan Island. Mataw fishers still go to sea in small rowing boats and use their traditional technology, whereas almost all sumuho fishers either now use motorized boats, or have abandoned the scoop net and switched to daytime fishing with drift nets, a technology introduced in the late-1980s.[3] Thus only the mataw fishers now continue fishing as prescribed by 'ancestral tradition'. Many other fishing techniques are used in Batanes, including fishing with nets, the 'flying net', diving with a spear gun, different kinds of hook-and-line methods, gill nets, and trawl nets (Hornedo 2000; Yamada 1967). Since the 1980s, the number of mechanized boats has increased greatly.[4]

Ivatans are famous for their resiliency toward typhoons, Batanes typically being affected by up to six a year between June and October (Blolong 1996). Given their isolation from the mainland and the natural hazards and challenges that must be faced, Tarrobago (2003:14) remarked that the Ivatans

> ... are a hardworking people who have learned to depend on themselves because they cannot count on help from the mainland, especially in months of bad weather. This sort of autonomy has also resulted in a culture of co-operatisation (*sic*). After a devastating typhoon, for example, the people of the community convene to take stock of the damage and work out a system of house repairs. All families have to send at least one member to help out. Voluntary communal labor is not limited to disaster response; it is also applicable to other community affairs such as house building, clearing of croplands, planting and harvesting.

That spirit of cooperation extends to the pre-existing traditions for the seasonal fisheries for Flying fish and Dorado, which functions in addition to manage near-shore fisheries.

The pre-existing coastal resource management system examined here is practiced by the fishers in Batan Island. However, the culture of fishing and the traditional seasonal activities and practices are shared among the islands in this area of the 'Bashiic' languages (Yamada 1997). Nevertheless in many areas there are local variations, and transformation and extinction of the traditional ways of fishing has occurred. This culture area also includes Pongso no Tao (the island of Lanyu, also known as Orchid Island or Botel Tobago, across the Bashi Channel and a part of Taiwan), whose inhabitants are said to have come from Batanes.[5]

---

[2]Until motors were the introduced in the 1980s, these were rowed.

[3]In the 1990s, large numbers of Flying fish would be caught and the price would fall dramatically at the height of the season.

[4]This observation is based on the work being done in a boat builder's workshop in Basco, in 1997, most of which was to convert rowing boats to accommodate inboard motors.

[5]According to comparisons of genetic markers the Yami of Lanyu are closer to the people of central Philippines than to the indigenous people of Taiwan (Datar 1999).

In Batanes from March until May mataw fishers focus on Flying fish (Exocoetidae) and the large golden Dorado (*Coryphaena hippurus*) or Dolphinfish (*Coryphaena hippurus*), two species of migratory pelagic fish that enter nearshore waters in significant numbers at this season. The arrival of these 'fish of summer' (*among nu rayon*) signals the start of the 'summer' (*rayon*) season. Attending closely to these signs and to the behavior of the fish, groups of fishers have long engaged in night fishing with lights for Flying fish, a technique called sumuho, or in angling for Dorado by day, a technique known as mataw. 'Mataw' refers to the specialized fishing method of catching Flying fish and subsequently using this as live bait for catching Dorado. The term 'mataw' also refers to fisherman who practice this method of fishing, which entails a commitment to go to sea to fish everyday for the entire three-month season.

Historical and ethnographic records (Gonzalez 1966) indicate that these seasonal fishing methods date back several centuries. And in recent years the mataw fishers have drawn on their status as users of traditional fishing methods to enforce regulations on their fishing grounds, in some cases gaining a measure of official recognition from local government as well as from sectoral organizations of fishers of the prior rights of the mataw vis-à-vis such competing new gears as 'drift nets' (gillnets). The mataw fishers of a vanua or 'port' regard their group as a cooperative institution, since it has formal associations, an elected leadership, 'laws', regulations, including sanctions, and prohibitions covering a clearly bounded season of fishing.

How the resource rights and management functions of the mataw traditions in Batanes Province operate is the focus of this chapter. Although the extent of pre-existing marine resource management systems is still being studied in the Philippines, it appears that the mataw system is one of the few that remains extant. Two other models are examined briefly in this chapter as comparisons; that of the Tagbanua people of Palawan Island, and the vastly different approach demonstrated by long-distance migrations of Visayan fishers. The mataw tradition comprises a third model. First I examine its ritual traditions and demonstrate how they function to form a fisheries social unit. This is followed by an analysis of the leadership and community regulations of the seasonal fishing activities. I conclude by examining change in this specialized fishery, focusing on gear type and marketing.

4.2 Indigenous or 'Pre-existing' Marine Resources Property Rights Regimes in the Philippines

In the 1990s, mataw fishing in Batanes seemed to be the only extant example of indigenous coastal resource management in the Philippines (Mangahas 1994). However, at a 2001 workshop on 'Marine and Coastal Resources and Community-Based Property Rights' it was observed that at least three types of local property rights regimes for marine and coastal resources are known to have been used in the Philippines among fishers and coastal communities

(Aguilar 2001).[6] All of those long antedate the community-based coastal resource management initiatives of local government, non-government organizations, or people's organizations, which began in the 1980s (Alcala 1996).

Those three models can be considered pre-existing, as they represent long-standing and even pre-colonial patterns of fishing and maritime movement in the archipelago. Generally, they denote operational informal laws and norms on access to resources that are not within the State's framework and rhetoric. Moreover, although the Local Government Code of 1991 provides for local legislative powers over municipal waters (defined as 15 km from the shoreline), as was noted in the same workshop, for most fisheries and fishing grounds the community dependent on the resources is not the same as the *barangay*.[7]

4.2.1 The Tagbanua Model

The celebrated case of the Tagbanua of Coron in northern Palawan represents one such model. For centuries, the indigenous Tagbanua made a living by diving for sea cucumber, and procuring valued commodities, like bird's nests. Tagbanua custom respected certain coral reefs regarded as the abode of large octopii and spirits, and which they were forbidden to approach (PAFID 2000). Also sacred are several lakes, to which they prohibit access by outsiders. They have also controlled and maintained clan rights over caves where bird's nests are gathered. Among their documented conservation practices is the passage of laws by the elders, such as that against gathering giant clams (*Tridacna sp.*) or cutting trees. The sanctions include corporal punishment (lashing) and use of bamboo stocks (Dalabajan 2001; Sampang 2005). Little had changed in the lives of the Tagbanua until mid-century: "Three factors explain how such equilibrium was made possible: a low population to resource base; an economy that functioned basically for subsistence and not exchange; and cultural norms that made it taboo to indiscriminately exploit the forest and coastal resource" (Dalabajan 2001:175).

In 1998, the Tagbanua claim for their Ancestral Domain, under the Indigenous People's Rights Act[8], including 22,400 ha of land and water (*teeb sorobleyen* or

[6]A fourth potential model represented by the customary practices of equity sharing among participants in the fish corrals in Bolinao, Pangasinan, which were awarded as fishing concessions by local government, is no longer in operation (S. Rodriguez, personal communication), see (Rodriguez 1997).

[7]The barangay is the smallest unit of local government in the Philippine system.

[8]This act passed in 1997 recognizes entitlements of 'indigenous peoples' in the Philippines over ancestral domain 'since time immemorial', including their rights to decide on the development of natural resources. It is seen as a milestone but also controversial, even among indigenous peoples and advocates of indigenous people's rights. (For a discussion of some dilemmas regarding the Tagbanua case, see Perez 2004.)

'inherited seas') was awarded by the Department of Environment and Natural Resources (PAFID 2000). However, this landmark achievement remains a continuing struggle, since the Tagbanua cannot fully monitor their large sea area against many threats. The problem is non-recognition by local government, and the proliferation of strangers: outside fishers, migrant families and their networks, and tourists and associated beach resorts. Powerful forces that wish to extract rent from the area are unhappy about the claim. The local government, which is not Tagbanua but dominated by migrants, feels excluded by a 'mere' cultural minority (see Sampang and Aguilar 2008; Perez 2004). Continued migration into some of the islands by fishers from the Visayas is facilitated by their social connections. Moreover, these new fishers (*dayo*) violate both the Tagbanua and national laws, aided by their supply network and market links that enable them to use such illegal fishing methods as blast fishing and sodium cyanide, as well as other destructive techniques.

4.2.2 The Visayan Fishers' Model

The Tagbanua model leads directly to the second model of a pre-existing pattern of access and rights to marine resources, which I term the 'Visayan fishers' model. This model emphasizes the mobility and interconnectedness of fishers across seas and islands. Migration is part of everyday life and an adaptive strategy as fishermen move and settle, following the marine resources that constitute their livelihood. It begins with a first small 'wave' of migrants who establish their base (*tumandok*) by settling on an island. Succeeding migrants or 'visitors' (*pangayaw* or *dayo*) may also settle more or less permanently in the same place. But newcomers, whether temporary visitors or settlers, cannot enter the area without first establishing a relationship with a local host. Such ties benefit both the host, who provides lodging and other facilities, and the sojourning fisher, by allowing access to resources (whether using small- or large-scale fishing gear), so long as there is no direct competition with the host (Zayas 1994; Palis 2001). A market for particular marine commodities motivates the sojourning fishers to explore or 'raid' (*pangayaw*) perceived frontiers, where the locals are either unaware of or lack access to newer fishing techniques that would enable them to harvest the same resource for themselves. Also probably they are not aware of the market value of these resources, or are unable to contact potential buyers and enter the market directly. The sojourning fishers return to their places of origin with either the end of the fishing season or of opportunities for profitable fishing.

Meantime, local fishers would have acquired some of the new fishing techniques from the 'visitors'. Eventually, migrant fishers would no longer be able to continue sojourning, either because they would be competing with the locals or as a result of resource depletion. They would have to move on, unless they decided to settle and integrate further with the community, or even switch to other livelihoods, like farming or animal husbandry. As resources become limited, exclusion would have set in, and relatively new settlers would have fewer rights and privileges than those with an 'established' status and therefore a longer-term commitment to the place. Only after one or two generations would the new migrants eventually have been able to shed their 'second class' status in the community.

In short, as Zayas (1994:126) puts it, sojourning fishers adopt 'maritime slash-and-burn' as a survival strategy. Population growth and poverty, together with novel fishing gears and the ability to cross seas and connect with other islands, impelled and continues to push Visayan fishers with limited opportunities in their places of origin to seek new frontiers. Basically, they gamble on establishing a new relationship with a host community, discovering new places to fish, and maneuvering to gain a niche against local fishers, at least in the short-term, by bringing in more advanced or specialized technologies.

This is the history of settlement of many islands, especially since the Second World War. It is a strategy that is vulnerable to overfishing and leads to rapid change in many fisheries. With increasing pressure on the fisheries, grounds become relocated increasingly seawards. Based on informal rules, nearshore fishing areas should be allocated for local subsistence needs, and worked only by small-scale fishers and always for the benefit of local residents. Areas farther seaward are perceived as where large-scale gears owned by the wealthy and powerful should operate, as should fishers coming from distant places. Along some coastlines, enclaves of fishers specialized in specific fishing methods may also develop and grow to enable a better support system, especially for those using more risky or illegal fishing techniques, such as compressors (Castillo 2009) or blast fishing (Galvez 1989).

In this fashion, Visayan fishers, especially those from Eastern Visayas or the islands of Cebu, Samar, Leyte, and Bohol, have moved around continuously, and extended their range to Mindanao, Palawan and Luzon, migrating from place-to-place to discover and extract as yet locally underutilized coastal resources. National and international demand for particular marine products, the exhaustion of a resource, and the quest for a living, keep fishers in constant motion and impel innovation, intensification and a high turnover of fishing gear technology.

4.2.3 The Mataw Fishers of Batanes Province

A third model of a pre-existing system of property rights and access in the Philippines is exemplified by the mataw fishing groups of Batanes. In contrast with the 'Visayan model', the context of fishing in Batanes differs by being protected by the dangerous waters of the Balintang Channel. This effectively excluded aggressive fishers from other parts of the Philippines.[9]

[9] However, there is competition with commercial fishers coming from Taiwan, who use more advanced technologies and large boats.

4.2.3.1 The Vanua as Meaningful Unit of Organization

The coastlines of the Batanes islands are fronted by either cliffs with shingle beaches and boulders, or by reef flats against which strong waves break (Photo 4.2). As a result, there are only a few places where boats can be safely launched and landed. Such places are called vanua in Ivatan, which the Ivatans translate into English as 'port'.

At present, four such ports are used by mataw fishers during the summer, and all are on the *valugan* side (approximately the eastern side) of Batan Island.[10] They have ancient names and attached stories and legends about ancestral fishers, including some about a mythical original fisher who introduced mataw fishing and tested all the vanuas of the island (Mangahas 2008a, b).

As the fishers point out, only four vanuas (referred to here as Chanpa-n, Manichit, Maratay, and Diora), among those around Batan Island still retain their traditional significance. They say that the rites to 'make the vanua' (*mayvanuvanua*), performed at the onset of the summer fishing season in the four vanuas, ensure that the fishers will be safe and the fishing good. This they contrast with the other vanuas, where fishers do not get together (or no longer do[11]) to perform rites, and

Photo 4.2 The coastal environment of a vanua, Batanes Province, Philippines

[10]People in Mahatao, Batan Island, orient themselves by at least four 'sides' of the island: *valugan*, *dichud* (meaning 'at the back of' Mt. Iraya, to the north of Batan Island), *kajbo* ('down below', south part of the island, where the Barangay of Imnajbu is found), and *kadpidan* ('the other side' or 'the side crossed over to', which is on the western side).

[11]Ritual practice died out for the vanua at Itbud (part of Uyugan) in the 1970s. Since that time the vanua has been modified by road construction, including removal of a venerated stone.

which are therefore regarded as being more 'accident-prone'. This, despite the sea on the eastern (*valugan*) side of Batan Island being somewhat rougher and more challenging during the summer. However, it is on that side of the island during the summer that many Flying fish are pursued by Dorado.

The towns of Basco, Mahatao, Ivana and Uyugan, on Batan Island, are located on the coast and close to river mouths, in the lowlands on the approximately the western side of the island. In the late-eighteenth century they were planned by the Dominicans and the people resettled (Hornedo 2000). So it may indeed be that the use of the four vanuas, which are all on the opposite side of the island, goes back through generations of fishers, as is believed by the present day fishers. There are some ancient settlements on the eastern side of the island, such as the archaeological site of Racuaydi (meaning 'large town') (Mijares 2001).

Thus, in addition to being places where fishers can access both land and sea, vanuas should also be appreciated as ancient or ancestral places. They are points in the landscape and seascape impregnated with the words and deeds of the ancestral fishers in times long passed, who used exactly the same places to go to and return from the sea. Being 'vanua of the ancestors', is expressed in the ritual words spoken during the rites before a fishing season begins.[12] This profound connection links the present day mataw community with very first fisher who performed the first sacrificial rites at the vanua and who left 'instructions' (*vidin*) on what is to be done each season. Each succeeding generation must either carefully reproduce the ritual words and acts of the ancestors, or risk grave misfortune, because if the spirits (*añitu*) are not appeased with the sacrificial offering they would contrive somehow to receive their 'share' of life. By performing the traditional rites the fishers sharing a vanua transform themselves into a collective, and embark on a socially complex power-laden negotiation with the fish and the spirits.

A vanua can have as few as five boats or as many as 30, as in Chanpa-n, the largest. However, the precise number fluctuates from year-to-year. A vanua is fairly open in terms of membership. For example, a fisher can stop fishing for a season or longer, or transfer to another vanua for various reasons, such as the need to be nearer his farm or to a temporary field house (*pañisanan*) in which he and his family would stay during the fishing season. Or a fisher from another island marrying into and settling in the area may also become a new member of the vanua. At this time they would realize that the practices are not the same across all vanuas. Some have different 'beliefs' (others none at all) and may 'make the vanua' slightly differently.

The vanua of fishers using traditional technologies to catch seasonal Dorado and Flying fish is also their home base and source of identification. Each fisher is identified at sea by the vanua to which he belongs. Mataw fishers keep watch to see which boats the fish 'are going to'. When they return with their catch they remark on who

[12]The words spoken by the lead fisherman during the rites at the beginning of the season explicitly invite the fish to a particular vanua: 'come to our vanua', the fish are called, it is 'the most beautiful vanua' (see Mangahas, 2008a, b).

attracted many fish to his boat, or who caught the fish that went to someone else, and similar comments. Over gin they relate how "... the fishers of Diora were catching many over there by the end of the bay, while the Lead Fisher of Maratay did not catch a single one ...", and so forth. As they compare each other's 'luck', there is a constant informal monitoring of individual fisher's success rates, as well as how the fishers identified with different vanuas are faring. Thus competition exists not only among fishers, but also among vanuas.

4.2.3.2 Inside the Vanua: Leadership, 'Laws' and Ritual Regulation of Seasonal Fishing Activities

The Lead Fisher is the focal point of a vanua as a group of fishers. As the first to go to sea and begin the season, the Lead Fisher has both a great responsibility to perform well and also the authority to do whatever he deems necessary to assure the collective success and welfare of the group throughout the season. The Lead Fisher is chosen based on his reputation as a 'good fisher' (Mangahas 2004). My informants explained that a Lead Fisher heads and is responsible for the vanua, just as the mayor is for a town, or as a 'king' for his followers and his kingdom, or as a father for his sons. Today the Lead Fisher of a vanua is called 'President', and the four vanuas on the eastern side of Batan Island have become formalized with written lists of the currently active members. An important officer is the Treasurer or Secretary, who maintains the records of dues paid by members. Some vanuas also have a 'Runner', or 'Information Officer', or a 'Sergeant-at-Arms', whose duty is to notify members of meetings and enforce orders from the President.

The main activity of the organization is the 'making of the vanua', at the beginning of the Summer fishing season. Just like members of a cooperative work project (*payuhwan*), the fishers of a vanua get together on an auspicious day (e.g., March 1) to 'construct the port' or 'make the vanua' before the fishing season can begin (Table 4.1). As in cooperative work, every member of the vanua should be present, or represented by a proxy if unable to attend. The actual 'work' is performing the sacrificial rite and distributing the meat of the sacrificed animal (a domesticated pig purchased specifically for this purpose) among those present.

Then on another carefully chosen date, the Lead Fisher makes the first fishing trip to 'inaugurate the vanua' (*umdinaw nu vanua*). The date is chosen based on the

Table 4.1 Sequence of ritual operations in *mataw* fishing

Operation	Meaning	Date
Mayvanuvanua	Making the port	March 1
Umdinaw nu Vanua	First fishing trip	March 5
Maynamunamu	Cleaning	April 14
Kapaychava nu Vanua	Dismantling the Vanua	After first week of May

pilaton, akin to an almanac listing the signs of the zodiac and auspicious dates.[13] Once these protocols have been observed and good portents seen, the other mataw fishers in the group follow in fishing. They are seriously committed to the activity, and fish for Dorado every day until the end of the season. By mid- or late-May, at a specific date (for example, May 15), the vanua is supposed to be 'dismantled' (*kavahen vanua*) by the Lead Fisher. By late-May the rainy season has set in, signaling the end of summer. Each fisher follows the Lead Fisher in scheduling the distribution of dried Dorado among his share partners (e.g. May 23) (Photo 4.3).

The image conveyed by the term 'making the vanua' is that of making, building or constructing something. However, this rite does not make any visible change in the landscape. Instead it sets the vanua apart as a sacred or sensitive place, where careless behavior is not appropriate. It is transformed into a kind of liminal or transitional area between land and sea, supercharged by taboos.

'Making the vanua' puts together the fishers as a cooperating unit ('*payuhwan*') for the duration of the season, or until the vanua is 'taken apart', during which the ritual speech would explicitly state that each fisher is now 'on his own'. For the mataw fishers, making the vanua is an integral part of the technology of fishing (Mangahas 2006) without which fishers believe they would have less success and also be vulnerable to misfortune. To fish is like inviting and taking part in hosting enigmatic visitors who come from afar and who are very sensitive (they easily 'get offended').

Photo 4.3 Distributing the catch to a landowner at the end of the season, Batanes Province, Philippines

[13]The book is copied by hand in ordinary notebooks and consulted often.

Meanwhile, the ancestral spirits expect 'payment' for the vanua, which is therefore given in advance of the season during the 'making of the vanua'. Observing the ritual contract calls for cooperation and conformity among the fishers so that the fish and spirits will favor their vanua with 'good luck'.

During the summer fishing season, traditional prohibitions (*dagen*) regulate certain types of behavior. These dictate the etiquette for the proper way of fishing, of handling the catch, of eating or distributing the catch and time when it should be done, among other things. This etiquette is followed by the fishers, those closely associated with them and visitors to the vanua during the fishing season. The objective of these prohibitions is to maintain social harmony, order and cooperation so the vanua can be made and kept 'clean and attractive to the fish'.[14]

Some of these prohibitions are relevant to marine resource management, as they may relate to equity or to the sharing of opportunities, whereas others may promote conservation indirectly. For example, one of the traditional prohibitions was an individual catch quota. Formerly, mataw fishers were limited to taking nine Dorado per trip. Should a fisher have already caught his quota yet wish to continue fishing, he should first return to shore to unload his catch before resuming fishing.[15]

Often controversial for the fishers wishing to use new gear types in the area are prohibitions giving exclusive access rights during the season to the fishers using traditional methods of mataw and *sumuho*. After the 'making of the vanua' it is forbidden to fish for demersal species using hooks and lines with sinkers, or to dive and fish with spear guns. Even swimming in the vanua or gathering shellfish along the shore is forbidden. All of these effectively implement an extended closed season for all fishing, except for the 'fish of summer'.[16] This is said to have had a beneficial effect on other marine resources, which had more time to grow and also regenerate. For example, lobsters caught in the vanua after the summer season were observed by fishers to have grown "quite large".

The logic of these prohibitions is that should fishers start paying more attention to other species after the seasonal fish have been called and invited to the vanua, then the seasonal fish might go away. If at mid-season the catch rate of many fishers shows a marked decline, it is thought to indicate that the vanua may have become 'dirty' because some prohibitions had not been followed. Fishers might overhear how a particular person was seen fishing improperly at the beginning of the season, or friends of a mataw fisher experiencing bad luck might wonder why bad luck occurs every time a particular visitor arrives. After informal discussions it might be agreed that traditional prohibitions have been violated. As penance a fisher may

[14]Related to this is that envy, resentment, arrogance, and non-cooperativeness are regarded as negative emotions and attitudes that also can affect fishing adversely.

[15]It seems that the quota was rarely attained, since I never saw more than seven Dorado caught during one fishing trip.

[16]Or practically half the year, from mid-October to mid-May, would be the off season, since diving is not usually done during the 'winter'.

respond to the social pressure of being perceived 'guilty' of transgressions by not fishing for a few days. Persons generally perceived to have 'upset the proper order of things' are advised informally to cease what they are doing. The Lead Fisher must perform a cleansing rite for the vanua. With that, collective anxiety is finally dispelled, and normalcy returns.

Even fishers using other gears are aware of the Mataws' normally unwritten 'laws' (*abtas*). However, some, like that at the vanua of Manichit, are written. There they possess a document written in Ivatan in 1940, and signed by the membership. Fishers in the vanua attached their signatures to a revised version prepared in 1960. The document states that the date for 'dismantling the vanua' will be determined by the Lead Fisher. The laws include penalties for failure to attend meetings, for those who steal or tamper with other fishers' gear, or who do not respect the date for 'dismantling the vanua'. Although other vanuas lack written rules, the traditional prohibitions are generally consistent and well known across the vanuas. Sometimes, modification of the rules can be made as appropriate to changed conditions, based on the previous season' experiences. This is done during the 'making of the vanua'. This could be regarded as sanctifying the new rules by presenting them to the ancestral spirits for acceptance. The test is whether the season turns out to be marked by success or misfortune.

The regulatory aspects of ritual can be seen as a vital element of the pre-existing system in terms of resource management. Such functions are apparent today, as they are the source of tension between users of different gears types. Recently, some prohibitions became the nexus of serious gear conflicts. Such conflicts sometimes reach beyond the vanua, and may be then resolved in larger political arenas. An example is the way in which one of the four vanuas became organized as a chapter of a sectoral organization of the fishers of Basco. In 1989 the Valugan Port Chapter of the Basco Fishermen–Farmers' Association (BFFA) passed a formal resolution to ban the use of drift nets in Valugan Bay, and to uphold the authority of the Lead Fisher. The resolution stated that fishers must "follow all instructions, or directions given or made by the Lead Fisher who was designated to make the first fishing trip (*mandinaw no vanua*) pursuant to traditional fishing practices in the area". The excerpts of the minutes signed by the President of the Valugan Port Chapter and the President of the Basco Fishermen–Farmers' Association BFFA comprise an interesting record of the negotiation between traditional and novel technologies and the final gear conflict resolution (Mangahas 2006) (Box 4.1).

The prohibited area covered by this resolution includes not only the vanua passage but also the entire fishing space enclosed by prominent points at either end of the bay, and even 'beyond' (according to rule #2). Essentially, it applies to all the fishers of the vanua Chanpa-n, wherever they fish.[17]

[17]Chanpa-n, facing Valugan Bay, is the largest *vanua* in terms of numbers of fishers, and the boats include traditional small *tatayas* as well as larger motorized boats. Mananioy Bay is the fishing ground for fishers from three vanuas (Manichit, Maratay, Diora).

Box 4.1 A Resolution Prescribing Rules and Regulations Governing Fishing Operations within the Tudaw-Achip Fishing Grounds at Valugan, Basco, Batanes, and Prescribing Penalties for Violation Thereof: Excerpt from the Minutes of Basco Fishermen–Farmers' Association meeting held on March 12, 1989 at Port Valugan (Chan-paan)

WHEREAS the Basco Fishermen–Farmers' Association is committed to preserve harmony among all fishermen fishing in the Rudaw-Achip fishing grounds and thus maintain peace and unity conducive to progress and develoment; and WHEREAS it has been observed that there are some fishermen who disregard the rights and welfare of other fishermen most particularly those engaged in dorado fishing;

NOW THEREFORE, be it resolved as it is hereby bodily resolved;

1. That no fisherman or group of fishermen are allowed to catch flying fish with nets in areas where other fishermen particularly the "mataw" are catching flying fish for dorado (arayo) bait within the areas between Rudaw and Achip.
2. That no fisherman or group of fishermen are allowed to fish with nets beyond the area designated by the group/association before May 15 of every year. Any person found violating this regulation shall be penalized with a fine of one hundred (100.00) pesos.
3. That all fishermen fishing in the area shall follow all instructions, or directions given or made by the leading fisherman who was designated to make the first fishing trip (mandinaw no vanua) pursuant to traditional fishing practices in the area.
4. That any person caught or found vandalizing any fishing banca, banca accessories and other fishing gears or equipments shall be penalized by a fine of one hundred (100.00) pesos or to change the damaged equipment or both fine or changing of the damaged equipment at the discretion of the BFFA officers.

(sgd.) President, Valugan Port Chapter
(sgd.) President, BFFA

In 1993 there occurred in one vanua a serious case of theft of a gillnet and sabotage of a mataw fishing boat. The gillnet was owned by an outsider, but it was being used by a recognized Lead Fisher of the vanua, who was fishing with the only boat of its size in the Bay that season. Emotion and anger ran high among fishers of both his vanua and others. The situation was also brought to the attention of the police, and was finally resolved through a Municipal Ordinance passed by the local government of Mahatao to 'regulate the preservation of cultural and traditional methods of capturing Dorado and other migratory fishes within the municipal waters of Mananioy Bay' (Box 4.2) (Mangahas 2006).

Box 4.2 Excerpts from Ordinance No. 03-03, Regulatory Ordinance for the Preservation of Cultural and Traditional Method of Fishing during the Months of March, April and May

"Sec. 3. It is strictly prohibited for any "matao" to use gill nets or any method other than the traditional way of catching flying fishes which are being used as baits for the migratory dorados on both sides of the restricted area indicated herein.

Sec. 4. Any fishers aside from "mataos" are prohibited to cast their gill nets intended for flying fishes inside the area herein described from Dispo Creek running perpendicular to an intersecting area between Mangavato and Pandangan pts. Gill net restrictions on this area shall be from the month of March, April and May. All other months are not covered by this restriction.

Sec. 5. Penal provision – Violation of this ordinance shall be subject to the following:

(a) First offense – P500.00
(b) Second offense – P1,000.00
(c) Third offense – P2,500.00 or imprisonment of 3 months upon discretion of the court

These records demonstrate concrete attempts to codify the traditional regulations in legal forms that would also be recognized and upheld by larger political structures, including the State[18]. It can be seen that the principles of equitable access to resources and of pre-existing customary rights are invoked in order to secure the limited access rule.

However, in practice the pre-existing system is also open to modification and compromise, to the point of potentially having no value for resource management. Although remaining consistent with the traditional form, fishers can be very creative in instituting new policies that can radically subvert the original intention, reflecting the changing concerns and priorities of the membership. For example, simply modifying the words spoken in 'making the vanua' can legitimize the use of other non-traditional gears for catching the 'fish of summer'. Similarly, the leadership structure can be reorganized. An example of that is provided by the vanua of Chanpa-n (also known as the Valugan Port Chapter of the Basco Fishermen–Farmers' Association). By 1997 (10 years after the earlier resolution, described above), driftnet fishing was allowed at the vanua. The vanua's 'Lead Fisher for Flying fish' now represented both the traditional sumuho fishers and the new drift net fishers. This Leader would be going to sea during the daytime instead of at night (i.e., he

[18]Municipal governments have this power under the Local Government Code of 1991 and the Fisheries Code of 1998 (Republic Act 8850).

would be using drift nets). After using drift nets, these fishers on their motorized boats would proceed to trolling for Dorado, thereby competing directly with the mataw fishers using oar-and-sail-powered smaller boats. As a compromise arrangement the drift net fishers were to give free bait to the mataw fisheries. The changes were probably reached with many of the vanua membership themselves also being interested in investing in and making use of new gears. If such a technology innovator was chosen as Lead Fisher, then he would have a powerful mandate to institute precedents on the fishing grounds.

The problem with modern technologies is that they cannot simply be set aside and discontinued, even if the riskiness of the undertaking is proven by experience. For example, many fishers from Chanpa-n remarked that the Flying fish catch has declined markedly in recent years, since many fishers switched to drift nets.[19] However, those who switched to drift netting had no option but to commit to the shift to try to recoup their investment in mechanization and gear, despite diminished returns.

In conjunction with this many fishers are becoming market-oriented, with a market for the catch especially among salaried employees and visitors in Basco. The seasonal fish is eyed for the monetary income that it represents, instead of a traditional class of rare goods that have value as a subsistence food as well as currency for many kinds of exchanges. The trend toward modernizing gear technologies and commercialization comes directly into conflict with the traditional prohibitions connected to the regulation of activities within the seasons, specifically the taboos against selling the catch during the season. Such prohibitions are explicitly in opposition to the market – forbidding sale of the substantial part of the catch for the duration of the fishing season. For those engaged in traditional mataw fishing, the real context of their engagement with the 'fish of summer' is a community economy founded on the value of dried fillets of Dorado as currency.

4.2.3.3 The Fishing Schedule and the Community Economy

The Dominican missionaries who came to Batanes centuries ago complained that they could not buy fish during the summer: 'There is very little fish and that is seasonal, and they did not want to sell it to us *because they believe that if they gave us fresh fish they could not catch more fish* so that the only fish we obtained was so dry and smoked that it was very difficult to eat. This is what follows from having been sent by God to a land characterized by an abundance of stones ...' (Llorente 1983:200, emphasis added). This demonstrates the time depth of the practices of seasonal fishing. It is noteworthy that the Dominicans' complaint also states the

[19]The first drift nets were introduced in 1987, and large catches were being made when I was in the field in the summer of 1992. At that time still relatively few motorized boats were using drift nets. However, drift net catches had declined significantly when I returned in 1997).

explicit reason why they could not buy fresh fish: were the fishers to sell the fish of summer 'they could not catch more fish'.

The maximization of cash income from market sales is not the economic incentive for mataw fishers. Rather, it is the value of Dorado to pay for many kinds of arrangements in support of other livelihood activities of the fisherman–farmer and his household. The seasonally abundant Dorado should not be sold prior to the end of the season, because the mataw fishers have entered into contracts for shares of the entire season's catch, and such contracts cannot be fulfilled until the end of the season. Hence the customary prohibition on consuming freshly caught fish is appropriate. Moreover, Dorado is also an item of reciprocity and sharing. Traditionally among mataw fishers only some parts of a fresh Dorado can be consumed immediately, shared or sold.[20] However, the main pieces, i.e. the fillets, must be dried in a specific way. But even the fillets should not be moved about before the end of the season.[21] Dried and smoked as they hang over the mataw's kitchen hearth, fillets are accumulated until the end of the season, and then distributed among the share partners during a special the sharing-out event (*payatay*). This occasion demonstrates how the Dorado fillets are used as a 'community currency' (DeMeulenaere and Lietaer 2003). For example, one day's field labor is equivalent in value to one dried Dorado fillet, and that is how mataw fishers would contract labor for weeding their fields during the summer fishing season. According to pre-arranged shares contracts, a 'share' of Dorado could pay for using a boat, or (if the mataw owns his own boat) for plowing and preparing a field for planting. Or it could also be exchanged for a set amount of cash given by the share partner before the season begins (like an investment). Nowadays some groups of mataw fishers use more complex sharing arrangements than in former times. The shares system has become more elaborate, notably among the mataws in Mahatao, where catch shares can be exchanged for use of land (Mangahas 2003).

For all those arrangements the fishing schedule also formally coordinates the proper time for consumption and distribution of the catch. The prohibitions ensure that the obligations of fishers to honor exchange contracts for their catch, to reciprocate favors using it, and to celebrate and share within their social network, are met before the fish can become a commodity and marketed. In this way the local economy based on the circulation of a locally produced value is protected, and becomes incorporated into an entire season as the emphasized unit of time.

[20]These include 'eggs' (*pya*), 'liver' (*atay*), 'stomach' (*vitnel*), thinned flesh from the fillet (*hathat*), *sindang* or a strip of dark flesh from the center of the fillet, 'ear' (*tadiña*), and 'bones'. Such pieces are the daily fare of the mataws and their households during the fishing season. They are also sent to share partners and given to friends. If not consumed, all can also be dried, except the fresh *hathat*, which is usually consumed raw as *lataven* (ceviche) immediately after fishing.

[21]Other traditional injunctions are that the fish cannot be put in a bag or a box (the only way of carrying Dorado is by a bamboo pole balanced on the shoulder). Neither can it be put in a bottle, sent by airplane, or loaded on any vehicle with wheels until the season is over, lest the fish become 'offended' and go away.

Filleted and dried, Dorado has great exchange value. Many would consider the catch too valuable to be consumed immediately. Dried Dorado is not only an edible item, but something appropriate for sharing with kin and friends; it has traditional value as a uniquely Ivatan product that ideally should flow to as many relevant people as possible (as one woman expressed it to me, 'so that they can also taste some Dorado'). Further, as something with a relatively long 'shelf life' (until the next summer fishing season), it has symbolic value as an item of stored wealth ('kept in a box above the hearth'), and also connotes subsistence and food security (as with the yam, *uvi*). In short, it stands for much more than a mere commodity.

In Mahatao, during the late-1990s, on rare occasions some mataw fishers either gave away or sold some of their fresh catch. However, it was done discretely, and then only to selected persons who would be careful not to display the gift or purchase. In contrast, in Basco it is apparent that the customary rule is not always being observed, because some fishers openly sold their catch immediately after landing it. The mataw fishers could do this because they had made different production arrangements, such as buying their bait instead of contracting to exchange a share of Dorado. For example, it could be that these fishers relied on cash income from either farming or remittances instead of contracting with share partners to organize inputs and livelihood arrangements during the fishing for the season. In theory, they could act increasingly individualistically. Without share obligations to fulfill, they would not need to fish continuously for the duration of the season, and could fish intermittently or as 'weekend mataws', instead of 'genuine mataws', who are committed to fish daily for the entire season.

The creation of wealth in the form of dried fish accrues to individual mataw fishers depending on their 'luck'. However, it is also a collective project. The 'fish of summer' are enticed to come to the vanua by the solidarity of the community, expressed as the 'cleanness' of the vanua. These enigmatic fish apparently favor particular individuals, and through time some mataws have become recognized as 'master fishers' (*sagal*) through their ability to catch many fish. One who is *masagal*, or a good fisher, has many good qualities, being 'popular' with fish as well as people, is characteristically generous, and shows leadership potential. Naturally, the person chosen to be Lead Fisher must be such a person, good in fishing as well as a man for others. This is a fisher suited to going from the vanua to call and seek favor with fish and spirits (Mangahas 2003). The group chooses its leader for his personal qualities that would be likely to attract good fortune and ensure the safety of the membership. Within this total economy the ecological units are not individual fishers, but collectives of organized and 'cooperating' fishers – i.e. the vanuas – engaged more in 'negotiating with' than 'extracting resources from' nature to ensure a continuing living for the entire community.

4.3 Conclusion

Three pre-existing systems of marine property rights or privileges in the Philippines have been identified and described briefly in the chapter. However, whether or not these models provide examples of resource management for the long-term is another question.

There are problems of enforcement, and generally the models are oriented more toward rights to subsistence, livelihood and equity of access than toward management. The idea of subsistence and the right to survive is embedded in all three models.

Population growth, migration, technology, the commoditization of nature and access to markets are among the principal driving forces that determine whether any pre-existing system would be stable or dynamic, or perhaps a tragedy waiting to happen. An extractive cycle such as that historically employed among Visayan fishers could not be sustained, as the escalation represented by an ever more rapid innovation and expansion of fishing effort inevitably leads to exhaustion of a resource and therefore to its collapse.

The Tagbanua have engaged in fishing for trade for many generations, and developed a customary legal system over their ancestral territory and resources. The Visayan strategy of mobility is one of riding on change and being at its forefront, based on a search for new resources, with the awareness that they will inevitably be exhausted and could only exploited by them (as outsiders) while frontiers remain unregulated.

Owing to geographical constraints Batanes is a fishery for insiders only. Very high fisher mobility and population pressures are not significant factors, unlike other parts of the Philippines. However, Batanes Province is not immune to technological intensification and market-oriented extraction focusing on the short- over the long-term. Nevertheless, the capture of Dorado and Flying fish within a traditional mode of livelihood linked with particular parts of the landscape, and with the steps, words and actions of the ancestors, remains strong in parts of the province.

The Dorado and Flying fish resources that are the focus in the regulation of summer fishing in Batanes are actually migratory species that cannot be 'managed' at this level. The seasonal traditions should therefore be appreciated as having an unintended impact on other fisheries, such as those for demersal species. The seasonal rites and prohibitions may serve to coordinate activities in complex adaptive fashion; closed seasons, fishing quotas, protected areas, and control over the activity of other gears are among the recognizable methods used. Other factors may underlie the continuing existence of the system of marine resource rights and management among mataw fishers in Batanes and account for its continued viability in some parts of the province (specifically the east side of Batan Island).

Built into the fishing schedule are the priorities of the traditional system. By making clear demarcations of time, space and people, the rite to 'make the vanua' initiates a cooperative approach to fishing. As a technological system, mataw fishing turns individual fishing into a collective activity in a manner entirely consistent with other traditional cooperative institutions in Ivatan culture. Following tradition, each fishing group of four vanuas on the east side of the island begins and delimits the season with a series of rites that regulate the behavior of people in relation to their landing sites and fishing grounds. The orientation of individual fishers is firstly toward sustaining other livelihood activities, rather than a market context. The system is based on cooperation rather than competition, with 'cooperation' referring to the observance of traditional prohibitions. In the cultural construction of fishing, 'the fish of summer' would then empathize with the human condition and give themselves to the group of fishers from the 'clean' vanua. Like the Tagbanua regard for their landscape with its forbidden areas, the traditional subsistence activities in Batanes invest their landscape with power. However, it should be noted in the modern context this is also seen as a fading and even obsolete power, from the perspective of those fishers interested in new technologies.

Vanuas are decentralized and run on democratic principles. As organized groups these have successfully resolved both local and external gear conflicts. By invoking tradition, some groups have managed to hold out and defend their fishing grounds against other forms of fishing during the summer (in Mahatao and in Basco). Some laws have been formally codified in the process. However, there is still much flexibility and creativity in the making of vanua policies.

Lastly, there is more than one currency in Batanes. The value of Dorado in exchange means that fishers have more autonomy to be self-sufficient and to make production arrangements within a community economy, apart from or complementary to the market economy.

Acknowledgements Research in Batanes in October-November 1991 and in the summer of 1992 was supported by a research grant from the Center for Integrative and Development Studies, University of the Philippines, Diliman. A pre-doctoral grant from the Wenner-Gren Foundation for Anthropological Research supported fieldwork in Batanes in the Summer of 1997.

References

Aguilar, R. (2001, June). *Ang Karagatang Ninuno ng Tagbanua-Calamianen: Isang Ulat at Pagpapakilala (The ancestral sea of the Calamian-Tagbanua: A Report and Introduction)*. Paper presented at the Marine and coastal resources and community-based property rights: A Philippine workshop, Anilao, Mabini and Batangas.

Alcala, A. C. (1996). Foreword. In E. M. Ferrer, L. Polotan-dela Cruz, & M. Agoncillo-Domingo (Eds.), *Seeds of hope: A collection of case studies on community-based coastal resources management in the Philippines* (pp. i–ii). Quezon City: College of Social Work and Community Development, University of the Philippines, and NGO Technical Working Group for Fisheries Reform and Advocacy.

Aprieto, V. L. (1995). *Philippine tuna fisheries: Yellowfin and skipjack*. Quezon City: University of the Philippines Press.

Blolong, R. R. (1996). The Ivatan cultural adaptation to typhoons: A portrait of a self-reliant community from the indigenous development perspective. *Aghamtao: Journal of the Anthropological Association of the Philippines, 8*, 13–24.

Castillo, R. C. (2009). *Way laing panginabuhi (There is no other livelihood): Negotiating danger and survival in the life world of a community of compressor fishers*. M.A. Thesis, University of the Philippines-Diliman.

Dalabajan, D. (2001). The healing of a Tagbanua ancestral homeland. In E. M. Ferrer, L. Polotan-dela Cruz, & G. F. Newkirk (Eds.), *Hope takes root: Community-based coastal resources management stories from Southeast Asia* (pp. 169–193). Quezon City: CBCRM Resource Center.

Datar, F. A. (1999). *Uvi: Ang Bigas ng Ivatan, Hanggang Kailan?* Professorial Chair Paper. CSSP Publications. Quezon City: College of Social Sciences and Philosophy, University of the Philippines-Diliman.

DeMeulenaere, S., & Lietaer, B. (2003). Sustaining cultural vitality in a globalizing world: The Balinese example. *International Journal of Social Economics, 30*(9). http://www.appropriate-economics.org/materials/sustaining_cultural_vitality.pdf. Accessed 20 April 2009.

Galvez, R. E. (1989). *Blast fishing and government response in Lingayen Gulf.* Quezon City: Institute of Social Work and Community Development, University of the Philippines-Diliman.
Gonzalez, P. J. (1966). *The Batanes Islands*. Manila: University of Santo Tomas Press.
Hornedo, F. H. (2000). *Taming the wind: Ethno-cultural history on the Ivatan of the Batanes Islands*. Manila: University of Santo Tomas Publishing House.
Hsu, Y.-C. (1982). *Yami fishing practices: Migratory fish*. Taipei: Southern Materials Center Inc.
Llorente, A. M. M. (1983). *A blending of cultures: The Batanes 1686–1898*. Manila: Historical Conservation Society.
Mangahas, M. F. (1994). *Indigenous coastal resource management: The case of mataw fishing in Batanes*. U.P. Assessments on the State of the Nation Occasional Papers Series No.94-001. Quezon City: University of the Philippines, Center for Integrative and Development Studies.
Mangahas, M. F. (2003, Oct). *Luck and leadership: traditional mataw 'firstfishers' in Batanes as 'men of power'*. Paper presented at the Philippine Political Science Association Conference, Davao City.
Mangahas, M. F. (2004). Fishing and performing fair shares. *AghamTao, 10*, 51–81.
Mangahas, M. F. (2006, June). *'Gear conflicts' and changing seascapes in Batanes, Philippines*. Paper presented at the 11th Conference of the International Association for the Study of Common Property, Bali.
Mangahas, M. F. (2008a, Oct). *A history of mataw fishing in Batanes, Philippines*. Paper presented at the CAPAS-IIAS Workshop on "Timescapes of Islands in the Asia-Pacific Region: Environmental History and Time Horizons of Management", Taipei.
Mangahas, M. F. (2008b). Making the vanua: collective fishing technology in Batanes and an Austronesian archetype of society. *Philippine Studies, 56*(4), 379–412.
Mijares, A. S. B. (2001). Settlement in stones: The Racuaydi site. *National Museum Paper, 11*, 106–115.
PAFID (Philippine Association for Intercultural Development). (2000). Mapping the ancestral lands and waters of the Calamian Tagbanua of Coron, Northern Palawan. In P. L. Bennagen & A. Royo (Eds.), *Mapping the earth, mapping life*. Quezon City: Legal Rights and Natural Resources Center, Inc – Kasama sa Kalikasan (LRC-KSK/Friends of the Earth-Philippines).
Palis, M. P. (2001). *Tumandok/Pangayaw: Identity, power and resource utilization in a Visayan Island*. M.A. Thesis, Ateneo de Manila University.
Perez, P. (2004). An exercise in reflexivity. *AghamTao, 10*, 120–134.
Republic Act No. 8550. An Act Providing for the Development, Management and Conservation of the Fisheries and Aquatic Resources, Integrating All Laws Pertinent Thereto and for Other Purposes. (February 17, 1998) http://daweb.da.gov.ph/FishCode/ra8550a.html. Accessed 20 April 2009.
Rodriguez, S. (1997). *The Barangen fishing concession in Bolinao: An ethnographic study of a customary marine tenure system*. M.A. Thesis, University of the Philippines-Diliman.
Sampang, A. G. (2005). *Ethnoicthyology and conservation practices of the Calamian Tagbanua in Coron Island, Palawan, Philippines*. M.S. thesis, University of the Philippines-Los Baños.
Sampang, A, G., & Aguilar, R. (2008, Oct). *Sacred areas in Coron Island, Palawan: Lessons learned from the Calamian Tagbanwa*. Paper presented at the 30th UGAT National Conference, *"Performing Heritage"*, San Fernando, Pampanga.
Tarrobago, A. (2003). *Bringing ICT to farmers and fisherfolk of Batanes, Philippines*. ICT Education Case Study, Asian South Pacific Bureau of Adult Education. 2009:http://portal.unesco.org/education/es/file_download.php/d7ed63b4b7b26e52734bbbf19aebf9a0ICT+Philippines+Case+Study.pdf. Accessed 11 April.
Yamada, Y. (1967). Fishing economy of the Itbayat, Batanes, Philippines with special reference to its vocabulary. *Asian Studies, 5*(1), 137–219.
Yamada, Y. (1997). *A Bibliography of the Bashiic languages and cultures*. Quezon City: CSSP Publications/University of the Philippines.
Zayas, C. N. (1994). Pangayaw and tumandok in the maritime world of the Visayan Islands. In I. Ushijima & C. N. Zayas (Eds.), *Fishers of the Visayas* (pp. 75–127). Quezon City: CSSP Publications/University of the Philippines Press.

Chapter 5
Pre-existing Inland Fisheries Management in Thailand: The Case of the Lower Songkhram River Basin

Malasri Khumsri

Abstract The livelihoods and food security of many Thai rural communities depend on inland capture fisheries, which are characterized by multiple species, diverse habitats and complex ecosystems. Current fisheries management in Thailand can neither control levels of exploitation and illegal fishing, nor can it achieve an equitable sharing of resources, which have become degraded and the focus of serious conflict. The capture fisheries of the Lower Songkhram River Basin (LSRB) exemplify those of a large, species-rich, tropical river basin, characterized by various ecological assemblages to which fishers have adapted with a range of gear and fishing techniques. There fisheries resources are managed jointly by national and local institutions under a complex multiple property rights regime. Fisheries property rights and their role in local fisheries management in the LSRB are examined, and conflicts explained. Changes in property rights are illustrated with particular reference to the illegal but tolerated large-scale barrage fishery. Local fishers' perceptions of fisheries management issues and collective responsibility for management are examined.

Keywords Fishers' perceptions • Inland fisheries • Local institutions • Management • Property rights

5.1 Introduction

Fisheries management in Thailand has both pre-existing and 'conventional' components, and includes various fisheries management strategies based on the Fisheries Act, B.E. 2490 (1947). It is implemented mainly by rules regarding permitted fishing methods, seasonally prohibited methods, licensing, and protection of spawning grounds.

M. Khumsri (✉)
Amnatcharoen Inland Fisheries Station, Muang District, Amnatcharoen Province, 37000, Thailand
e-mail: kmalasri@yahoo.com

K. Ruddle and A. Satria (eds.), *Managing Coastal and Inland Waters: Pre-existing Aquatic Management Systems in Southeast Asia*, DOI 10.1007/978-90-481-9555-8_5,

Exclusive, secure property rights do not exist in Thai fisheries, which are considered an open access resource. However, many communities restrict open access, and allow fishing by residents only. This has led to territorial conflicts with newcomers.

Resource conflicts are a major issue in Thai inland capture fisheries. Some occur between resource users and government, others between downstream and upstream communities, and yet others among fishers with different interests. The causes vary. Use conflicts also occur when various types of fishing gear increase in number. In addition, when government officers try to enforce strictly the fisheries regulations pertaining to illegal gear, conflict between them and fishers often becomes serious, and now is addressed at the national level. For instance, numerous lawsuits over the violation of fisheries regulations have arisen in the Songkhram River Basin, resulting in conflict between patrolling Department of Fisheries (DoF) staff and fishers, as well as among fishers. This led local communities and local politicians to request the government solve the problem of the use of destructive fishing methods.

In the Lower Songkhram River Basin (LSRB) pre-existing systems of local resources management are based on property rights, which themselves require elucidation. In the LSRB rights to manage fisheries have various sources and are exercised differently. Based on a field study conducted in 2007, the fisheries property rights system and conflicts within it are explained. Changes in such property rights are illustrated with particular reference to the illegal but tolerated large-scale barrage fishery. Local fishers' perceptions of fisheries management issues and collective responsibility for management are examined.

5.1.1 The Lower Songkhram River Basin

The LSRB is located in the Lower part of the Mekong Basin (Fig. 5.1). It is the most fertile river basin in Thailand, and is the largest spawning ground of aquatic animals in the lower Mekong River basin. The floodplain of this basin is complex, comprising different wetland habitat types that include rivers, pools, tributary streams, seasonally flooded-forests, grasslands, swamps, reservoirs, and ponds. Approximately 39% of the basin area is used to cultivate rice and the remainder is under upland field crops, with only remnants of forest remaining (Blake 2006). About 54% of the lower basin catchment is wetland, including rice fields, which cover 108,000 ha during the June–October wet season (Blake 2006).

The complex physical characteristics and ecosystems of the LSRB result in a multifaceted and dynamic fishery that is not well-understood by outsiders. The natural complexity of the region resulted in diversified livelihood strategies that depend heavily on a combination of fishing, hunting and gathering in biologically diverse wetland environments for both household subsistence and economic purposes. Fishing in the LSRB ranges from a part-time, small-scale subsistence activity, to full-time, commercial, large-scale fishing. There is a high level of participation in fisheries throughout the basin, especially at the household level, with men, women and children of all ages involved.

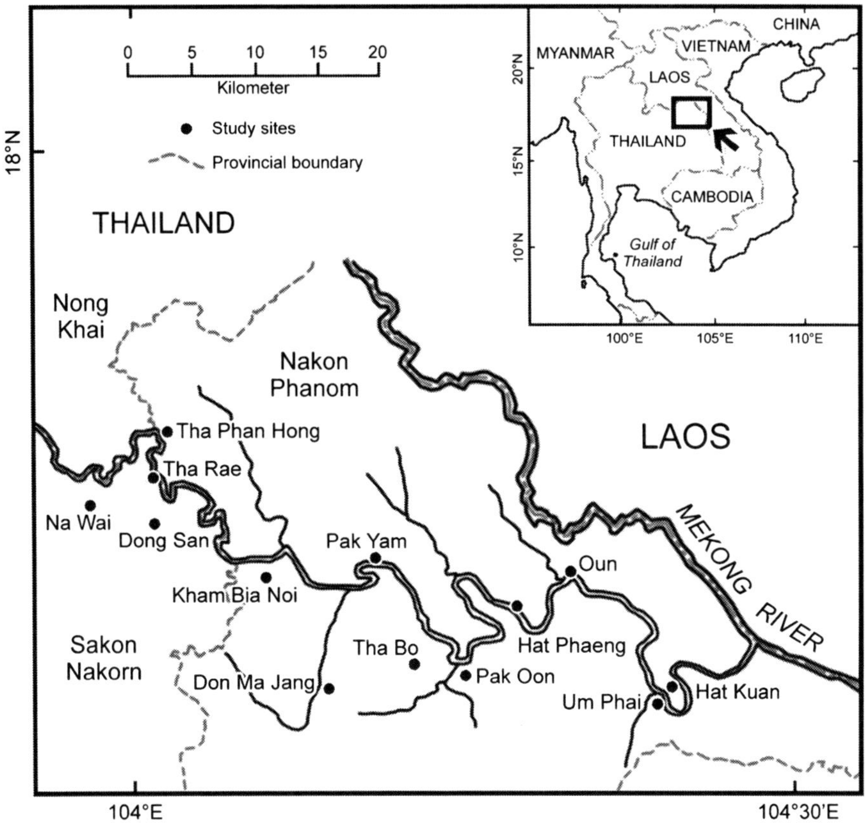

Fig. 5.1 Location of the study area

The main habitat of LSRB floodplains is the seasonally flooded forest, which is usually inundated for several months, and used by many native and migratory species of fish as a spawning and nursery ground. The water level recedes during the period September–December, at the end of rainy season. At that time fish are intercepted using a wide variety of fishing gears at such bottlenecks as lake exits, when they return to the mainstream. In the dry season the flooded-forest becomes a fragmented mosaic of various aquatic and terrestrial wetland habitats. These are harvested for diverse edible aquatic organisms and terrestrial foods, medicinal or households use items, fuel wood, and items of cultural importance, among others. Around July–August, as the rains start and the river level begins to rise again, villagers dig in the forest for bamboo shoots, and also collect a wide variety of plants and edible fungi for both family consumption and sale.

However, in recent years, most seasonally flooded forest in the Northeast Region of Thailand has been severely degraded and ecologically 'distorted' as a consequence of the multiple hydrological and geomorphologic changes to the river, and sediment flows. Although the remaining seasonally flooded forest in the LSRB is relatively the most intact of flooded-forests ecosystems remaining in the Northeast Region (DEQP 2002), nowadays it is faced with numerous threats. These include mainly logging and charcoal-making and conversion to either agricultural land or industrial plantations. Flooded-forest areas have been allocated to individuals, in line with government policy. This has resulted in conversion of the formerly common property to individually-owned cultivable areas, which the owners can invest in independently.

During the last 60 years the economy of the Songkhram River Basin has changed from subsistence to commercial operations, with state interventions to control resource use. Together these factors have resulted in overexploitation of natural resources and serious conflicts over their use. Further, the soils of the basin are mainly sandy with low levels of organic matter, which therefore are eroded easily through inappropriate agricultural practices. In addition, a saline soil in some parts of the LSRB limits rice production and the development of agriculture in general. Most former flooded-forest has now been converted for agricultural use; most upland zones are devoted to annual cash crops of cassava and sugarcane; and most floodplain areas are used for dry season rice cultivation, normally during the December–April period. This intensive dry season rice cultivation causes many problems, such as natural resource degradation, conflicts in natural resource use, risks to human health, and unnaturally high erosion rates from soil compaction, soil exposure and regular plowing. Soil loss caused by erosion in the LSRB was estimated at about 145,473 million tons per annum (DEDP 1997). Land use change has resulted in altered river and floodplains ecological systems, which often harms fish habitats. Many LSRB fishers claimed that most deep pools in the Songkhram River and its tributaries have gradually become shallower owing to sediment deposition on the lower floodplain. Increasing sediments at the mouth of the Songkhram River obstruct water flow, thereby affecting the annual flooding, especially the storage and steady release of floodwaters throughout the dry season. Development in the Songkhram River Basin has had both direct and indirect impacts on wetland resources and biodiversity. The continual external pressure to develop irrigation systems and domestic water supply has caused the degradation of natural resources. And the national policy to expand industrial tree plantations in the Upper Northeast Region has led to the rapid growth of rubber, eucalyptus and oil palm monocrop plantations, with the attendant ecological problems of lowered water tables, increased erosion, reduced soil fertility, and diminished local biodiversity.

5.1.2 Fisheries in the LSRB

Fishing plays an important economic role in the LSRB, with an estimated 63.28% of all households engaged in it either full- or part-time (MRC 2000). Fishers in the LSRB operate in rivers, reservoirs, community ponds, swamps, and rice fields. Usually, small-scale fishing gears are used in flooded areas, reservoirs and rice fields near a fisher's home. In contrast such large-scale gear as the stationary trawl bag net, large cylinder trap and barrage (Photos 5.1–5.5) are often placed where they encroach on neighboring communities, owing to their specific placement requirements. Most large-scale gear is used mainly in the Songkhram River and its tributaries.

The main capture fisheries in Northeast Thailand depend on the Mekong Basin, with the most important rivers for them being the Songkhram, Pong, Mun, and Chi. The River Songkhram is the most fertile, and has the highest biodiversity of any

Photo 5.1a Barrage size and configuration is adjusted to the conditions of a specific fishing territory

Photo 5.1b Barrage size and configuration is adjusted to the conditions of a specific fishing territory

Photo 5.1c Barrage size and configuration is adjusted to the conditions of a specific fishing territory

Photo 5.2 Details of barrage construction

Photo 5.3 Bamboo fence with a mosquito net installed to block a creek

Photo 5.4 Seine nets and electric shocks are used to harvest all fish after water flow has ceased

Photo 5.5 A barrage after an auctioned season has finished

freshwater habitat in Thailand (Boonyaratpalin et al. 2002). It supports a large capture fishery associated in particular with extensive wetlands, but where natural lakes, rice fields, reservoirs, and rivers are all fished (Hortle and Suntornratana 2008). A high of 183 species was recorded (OEPP 1999), and low figures are 53 (KKU 1997) and 32–70 species (Yingcharoen and Vilapat 2000). Boonyaratpalin et al. (2002) identified 149 species representing 33 families. The wide range reported reflects mainly either differences in methodology and sampling sites, or differences in annual flood levels. However, the fish diversity in the LSRB is higher than that of the Pong, Chi and Mun River Basin, where only 96 fish species representing 28 families were recorded (Sricharoendham et al. 1998) and that of the Tha Chin Basin, where 77 fish species representing 21 families Sricharoendham et al. (1998) were recorded. A connection with the Mekong River, which permits the ingress of Mekong species, might account for the high diversity in the LSRB (Rainboth 1996).

To ensure good catches, most fishers change their fishing grounds and territories seasonally, and they normally work different areas in open and closed seasons. Most (41%) normally fish in the flooding areas near their homes during the closed season, whereas about 37% fish in both the Songkhram River and the flooded areas during the open season. Because fishers have many fishing grounds available, owing to the large flooded area in the closed season, they can fish near their villages and in rice fields. However, fishing grounds are limited in the March–May dry season, when water remains only in the mainstreams. At that time fishers go far from their village, and normally operate in the main rivers and large reservoirs.

Fishing is done year-round in the LSRB, but the type of gear used varies by location and season. Gear usage also varies by the fish species targeted, and particularly according to their feeding behavior. Seasonality of gear use is related also to water level. During the May–September period, when water levels increase with flow from the Mekong River into the Songkhram River, fish migrate with the flow for spawning and early growth, and are caught mostly with hook, long line, small trap, gill net, cover pot, casting net, spear, small bag net, and tube trap. The main fishing season is from September to November, when water levels decrease with a return flow toward the Mekong River, and large numbers of fish leave the Songkhram system in a return migration toward it. They are caught in large quantities using the barrage, trawl bag net, beach seine, lift net, and brush parks, among other gear types. Finally, the December–April low water or dry fishing season is when such small-scale gear as the casting net, gill net, long line hook, and scoop net predominate (Table 5.1).

There are additional detailed variations. For example, although most small-scale fishing gear is used throughout the year, mesh-size changes to correspond with fishing time and sizes of fish targeted. Some gears can be used only in the daytime and others only at night, whereas other types, such as most passive or stationary gear, are used round-the-clock.

The type of gear used correlates also with the purpose for which fishing is done. Small-scale gear is used throughout the year, mainly to catch fish for home consumption. In contrast large-scale gear is used by commercial fishers mostly to catch fish during their return migration to the mainstream. For example, the barrage is used to block a creek or river using fences and nets to catch all sizes of fish during their return migration, from mid-October to January. It is normally used after fishing with the stationary trawl net has finished. The beach seine is effective when water levels are relatively low, from April to mid-June and mid-October to March. The vertical cylinder trap is used during the June–August and September–October seasons of high water. Push nets are used from October to November, when the water level remains at its highest.

In the LSRB the type of fishing gear used varies also according to season and the physical characteristics of the location fished (Table 5.2). The 22 gears commonly used in the LSRB are divisible into small- and large-scale types. The former are characterized by low catching rates, and are made from both modern (nylon) and traditional materials (simple materials such as wood and bamboo). Most gear is small-scale (83.2%). A total of 13 small-scale gear types were indentified, the main ones being the gill net (29.4%), long line hook (13.1%) and casting net (13%). Large-scale fishing gear has higher catch rates and normally requires larger investments. Only nine are classified as large-scale, the three main ones being the stationary trawl bag net (12.5%), barrage (2.6%) and surrounding seine net (2.2%). As fishing is most intensive in the middle basin, the largest percentage of gear is used there (53.4%), whereas the lowest percentage occurs in the lower basin (13.7%).

Gear usage differs among the three parts of the LSRB. Although most small-scale gear types are used throughout the basin, they are used much more in the lower part of basin, particularly in villages located within 1–2 km of the Songkhram River, and for fishing on floodplains. In contrast, large-scale gear is used mostly in the upper and middle parts of basin. Also, the use of large-scale gear types differs within the LRSB. The stationary trawl net is used mainly in the upper and middle parts, particularly in

Table 5.1 Gear use by season and fishing time

		Fishing season of different types of fishing gears used (month)												Fishing times	
No	Types of Fishing gears	January	February	March	April	May	June	July	August	September	October	November	December	Day	Night
1	Gill net	◄	—	—	—	—	—	—	—	—	—	—	►	/	/
2	Long-line hook	◄	—	—	—	—	—	—	—	—	—	—	►		/
3	Casting net	◄	—	—	—	—	►							/	
4	Vertical trap for small fish					◄	►			◄	►			/	/
5	Stationary trawl bag net									◄	►			/	/
6	Hook	◄	—	—	—	►									/
7	Scoop net	◄	—	—	—	—	—	—	—	—	—	—	►	/	
8	Cover pot	◄	—	—	—	—	—	—	—	—	—	—	►	/	
9	Eel Tube trap	◄	—	—	—	—	—	—	—	—	—	—	►		/
10	Spear	◄	—	—	—	—	—	—	—	—	—	—	►		/
11	Push net										◄	►		/	/
12	Surrounding seine net	◄	—	—	—	—	►				◄	—	►	/	
13	Small barrage								◄	—	►			/	/
14	Large vertical cylinder trap with large mesh-size					◄	►							/	/
15	Vase trap for *Hemibagrus nemurus*						◄	—	—	►					/
16	Large lift-net									◄	—	—	►	/	/
17	Large barrage	◄	►						◄	—	—	—	►	/	/
18	Large vertical cylinder trap with mosquito net						◄	—	—	—	►			/	/
19	Horizontal cylinder trap for Snakehead fish	◄	—	—	—	—	—	—	—	—	—	—	►	/	/
20	Small lift-net					◄	—	►						/	
21	Brush park		◄	—	—	►				◄	—	►		/	/
22	Frog trap										◄	►			/

Table 5.2 Common fishing gear used in the LSRB

		Percentage of fishing gears used in different parts of the LSRB (%)			
No	Fishing gear type	Upper	Middle	Lower	Total (%)
1	Gill net	8.4	13.4	7.6	29.4
2	Long-line hook	5.1	6.5	1.5	13.1
3	Casting net	6.5	4.8	1.7	13.0
4	Hook	2.1	2.5	0	4.6
5	Scoop net	0.4	2.4	0.6	3.3
6	Cover pot	0.7	2.2	0.3	3.2
7	Eel Tube trap	0.3	2.1	0.1	2.5
8	Spear	0	1.9	0.1	2.1
9	Vase trap for *Hemibagrus nemurus*	1.2	0.6	0	1.8
10	Horizontal cylinder traps for Snakehead fish	0.1	0.7	0	0.8
11	Small lift-net	0	0.1	0.3	0.4
12	Brush park	0	0.3	0	0.3
13	Frog Trap	0	0.1	0	0.1
	Total small scale fishing gear	26.5	43.7	13	83.2
1	Vertical trap for small fish	1.7	4.6	13	6.3
2	Stationary trawl bag net	1.7	4.6	0	6.2
3	Surrounding seine net	0.2	2.0	0	2.2
4	Push net	1.5	0.3	0.3	2.1
5	Small barrage	0.6	1.1	0.1	1.8
6	Large barrage	0.4	0.4	0	0.8
7	Large vertical cylinder trap with large mesh-size	1.2	0.4	0.3	1.9
8	Large vertical cylinder trap with mosquito nets	0.7	0.1	0	0.8
9	Large lift-net	0.1	0.8	0	0.9
22	Total Large-scale fishing gear	6.4	9.7	0.7	16.8
Total		32.9	53.4	13.7	100.0

villages near the Songkhram River, but not the lower part, since its requires a rapid and strong current. The barrage is operated only in the upper and middle parts. Beach seines are commonly used in the lower part, particularly at the river mouth.

The inherent complexity was heightened by the development of fishing techniques to meet the increasing demand for fish, as fishing techniques having been developed to maximize catches from various habitats. Further, the LSRB ecosystem has changed overtime, with changes in water level and current flow rates occurring at different seasons and in different location within the basin. This influences fish abundance and feeding behavior. As a result fishers have developed their gear types and methods to suit these physical and seasonal parameters. Economic growth has resulted in a change from subsistence to commercial fishing.

There have been three phases in the evolution of fishing gear in the LSRB. During the 'Early Period' (1893–1952) fishing operations were based on traditional and non-destructive fishing gear types, most of which were made from locally available natural materials. Most fish caught in those days were used either for home consumption or bartered. In the 'Middle Period' (1952–1977) modern materials like nylon nets replaced local material for making fishing gear, and boats were motorized. In the 'Modern Period' (1997 to present) large-scale gears like the stationary trawl net and barrage were introduced and adopted by the richer people, whereas the poorer people developed their fishing gear to include the large vertical cylinder trap and motorized boat.

Most fishers in the LSRB have been able to develop their fishing techniques to increase catches, such that most large-scale fishing gear is now illegal, according to the Fisheries Act, B.E. 2490 (1947). However, since the law cannot be fully enforced, such illegal fishing gear types are now used throughout the basin.

5.1.3 Occupation and Dependency on Fisheries

Fishing is a main occupation of 23% of respondents. This varies within the LSRB, from a high of 35.7% in the middle part of the basin, where, together with the upper part, most large-scale fishing is concentrated, to only 5.0% in the lower part, where fishers are commonly part-timers using small-scale gear. Khumsri et al. (2009) demonstrated the level of dependency on fisheries in the LSRB by an analysis of household occupations. The main occupations of most people (76.8%) are not related to fisheries, and of those 31.4% are agricultural and 45.4% non-agricultural (Table 5.3). Rice farming is the main occupation of 17.9% of those whose main occupation is agricultural. Migration of young people to work either in Bangkok or its vicinity was the highest of the main non-agricultural 'occupation' of 22.5% of the population surveyed, followed by 'general local worker'.

Table 5.3 Main occupations of fisheries households

	Main occupations in different parts of the LSRB (%)			
Main occupation	Upper part	Middle part	Lower part	Total (%)
Fishing	15.0	35.7	5.0	23.2
Non-fishing	85.0	63.6	88.3	76.8
Agriculture	40.0	22.1	33.3	31.4
Rice farming	25.0	13.6	18.3	17.9
Livestock	3.8	7.1	3.3	5.3
Water melon cultivation	8.8	0	0	2.5
Rubber plantation	2.5	1.4	11.7	3.9
Cage culture	0	0.7	6.7	1.8
Non-agricultural	45.0	41.4	55.0	45.4
Grocer	2.5	1.4	1.7	1.8
Fish trader	0	0	3.3	0.7
Community official/leaders	5.0	7.9	0	5.4
General local employee	16.3	7.1	30.0	14.6
Remittance	21.3	24.3	20.0	22.5
Services	0	0.7	0	0.4
Total	100.0	100.0	100.0	100

Numerous activities supplement household incomes. The main ones are fishing, rice farming and 'general local work', engaged in by 67.1%, 14.3% and 6.4% of respondents, respectively. Fishing has been is either a main or secondary sources of household income, along with the remittances from community members now working in cities. Most people also still collect wild foodstuffs for everyday use from the seasonally flooded forest.

5.1.4 *Fisheries Household Economics*

Average annual house income differs slightly between different categories of activities. The average annual household income was equivalent to US$1,832,[1] which is far above the poverty line of Thailand (>US$588 per year). However, the average household debt was US$1,684, which is very high. The average annual fishing household income was about US$400. This varied from zero, where fishing is only for home consumption, to US$4,400 per annum per household. In other words, the fishing income contributed about 21.7% to total annual household income, whereas the remaining 78.3% was derived from non-fishing activities (Khumsri et al. 2009).

The relationship between total annual household income and fishing income varies between groups of fishers, with that of large-scale fishers being significantly higher than that of small-scale fishers. These income differences are attributable to the relative effectiveness of gear use. In addition, there is a significant difference in fishing income among three different locations in the LSRB; fishers in the middle part having the highest (US$633) whereas the lowest income is found in the lower part of the basin (US$157). The higher income of the fishers in the middle part of the LSRB is attributable to the large fishing gears and to a longer fishing season (Table 5.4). In terms of annual household income and debt, the economic condition of large-scale fishers is better than that of small-scale fishers.

5.1.5 *Fishers' Perception of the Condition of Fisheries Resources*

Fishers' perceptions on trends of fish abundance during in the period 2005–2008 were examined (Khumsri et al. 2009). Although it has not been demonstrated scientifically that fish resources in the LSRB have decreased, most respondents (76%) claimed that a decline was evident. However, fish production in the LSRB depends on flood level; if it is high, fish production is also high. This was mentioned by 13% of fishers. That the fish productivity of the LSRB remained stable at a high level was confirmed by 9% of them.

---

[1] During the study period 1 USD equalled approximately 34 Thai baht.

Table 5.4 Annual income and debt of fisheries households between different groups of fishers and in different location of the LSRB

	Fisher category (USD/household/year)			Part of basin (USD/household/year)				
Averages income/debt	Large fishers	Small fishers	t-test	Upper	Middle	Lower	One-way ANOVA	Total
Total household income	2,728	1,610	4.540*	1,846	2,011	1,550	2.338	1,842
Fishing income	1,201	226	8.929*	294	633	157	18.831*	400
Non-fishing income	1,492	1,315	0.774	1,480	1,282	1,326	0.561	1,351
Agricultural	526	525	0.006	607	397	635	3.001	526
Non-agricultural sector	966	789	0.859	872	885	690	0.818	825
Total household debt	19,289	1,607	0.905	21,449	1,445	1,545	2.246	1,684

Note: *significant at 95% confidence level.

More than half the fishers (58%) in the middle area were satisfied with the present level of fish abundance. About 16% expressed high and very high levels of satisfaction, whereas a low and very low level of satisfaction was given by 26.4%. This indicates clearly that although it has been claimed that fish production in the LSRB has decreased, most fishers are still satisfied with the present level of fish abundance.

Human activities and resource use patterns often have a negative impact on fisheries resources. Fishers identified five key causes of fisheries resource degradation: (1) use of destructive fishing gear, (2) destruction of flooded-forests, (3) weir and irrigation development, (4) fishing during the spawning season, and (5) fishing during fish spawning migrations. Operation of such destructive gear as the barrage, stationary trawl bag net and seine net were regarded as the problem having the most serious impact on fisheries resources (Khumsri et al. 2009).

Khumsri et al. (2009) also examined fishers' perceptions of the management strategies for rehabilitation of fisheries resources. Management strategies for improvement of fisheries resources rated at the 'very high' levels were (1) the establishment of fish conservation zones, (2) a ban on destructive fishing gear, (3) the prohibition of fishing during the spawning season and migration, and (4) rehabilitation of flooded forests.

Throughout the LSRB fishers claim that the main problem is that they are scared of the DoF Patrol Officers, and that this restricts their freedom of fishing (40%). Loss of fishing gears was raised by 39% of fishers, and decrease in the profitability of fishing was mentioned by 35%. These problems were mostly found in the middle and upper parts of the basin, most likely because these are the major fishing areas.

The problems of large-scale fishers and small-scale fishers differed significantly. Major problems for the latter are the loss of fishing gear (36%) and the decreased profitability of fishing (22.9%). The fear of being arrested by Patrol Officers is the most serious problem for large-scale fishers (20.1%), because most fishing gear used by them is illegal.

5.2 Property Rights System in Fisheries Management in the LSRB

Understanding property rights systems is basic to understanding the local management of resources. At present, fisheries resources in the LSRB are managed concurrently by local communities, based on pre-existing or de facto rights, and de jure by the DoF, according to the Fisheries Law of 1947. Further, according to the Thai Civil and Commercial Law of 1925, natural resources used in common, such as shores, streams and lakes, are State Property (RTG 1930). However, at the same time local communities recognize that individuals have ownership of fishing rights in such areas, and that they also have the right to exclude others from fishing within them. The result is a complex and multiple set of overlapping, complementary and conflicting individual, common and state property rights within a single, small geographical area used as a fishing ground (Khumsri et al. 2009).

5.2.1 *Customary Rights over Fishing Grounds*

The choice of fishing in the LSRB depends on both season and the rights over suitable fishing grounds, dictated by the different operational requirements of the various fishing gears for a particular fishing ground. Communities in the LSRB recognize differing 'bundles' of de facto rights over fishing grounds, the ownership of which is restricted to those families, relatives or partners with traditionally established user rights over particular water bodies. The principal bundles of rights are (1) property rights as an authorized user, (2) property rights as a proprietor, and (3) property rights as an owner (Table 5.5).

An 'authorized user' has the de facto right to place small fish traps and longlines across watercourses. The first occupants of these fishing grounds at the beginning of each fishing season are recognized as the sole rights holders for that season only, and others are not permitted access. These are simply operational rights for authorized users to access and catch fish in designated areas, and do not allow participation in collective action to determine operational rules for harvesting or exclusion, which are defined by local community members, based on custom.

Communities recognize as a 'proprietor' those of their membership who own large vertical cylinder traps, seine nets and large lift nets, all relatively efficient gear, the effectiveness of which depends mainly on fishing location. Because of their large size, these gear types usually require permanent installations on a dedicated patch of land. In general, those recognized as proprietors are first occupants. The most suitable grounds for these gear types are all owned and fished each year continuously by the same proprietor. Since these rights are generally inviolable, locations available to newcomers are rare, except when proprietors do not exercise them for one or two years. Proprietors can transfer the rights to their children or other relatives, but rights cannot be sold.

Table 5.5 De facto rights to fishing grounds in the LSRB by gear type and status of user

	Type of right				
Gear Type	Access and withdrawal	Management	Exclusion	Alienation	Status of user
Small fish trap	×	–	–	–	Authorized user
Long line	×	–	–	–	Authorized user
Large vertical cylinder trap	×	×	×	–	Proprietor
Seine net	×	×	×	–	Proprietor
Brush park	×	×	×	–	Proprietor
Large lift-net	×	×	×	–	Proprietor
Stationary trawl bag net	×	×	×	×	Owner
Barrage	×	×	×	×	Owner

Adapted from Ostrom and Schlager (1996).
Note: × indicates right possessed.

Rights holders of grounds for stationary trawl bag nets and barrages are regarded as 'owners' of fishing grounds that are just like plots of rice land. Because these two large gears target fish during their return migration to mainstreams, their proper placement is the main determinant of harvesting rates. As a consequence, all the best locations have long been owned. The basic features of these rights are that (1) owners can exclude others from their fishing ground, and (2) the rights can be sold, rented or inherited.

5.2.2 Returning Rights from Private to Common Property in Barrage Fishing

As has been widely noted (e.g., Ruddle 1994), economic, political and related change triggers an alteration of property rights regimes. This has occurred throughout Thailand since the late-1950s, as the rural economy changed from local subsistence and barter to external market-oriented commercialism. Then, in the 1980s, the political system was decentralized, and Sub-district Councils and Administrative Organizations (*Or-Bor-Tor*) were authorized to manage natural resources (Khumsri et al. n.d.).

The barrage fishery is the most lucrative commercial fishing gear used in the LSRB, with an annual income ranging from US$1,516 to 31,513, and catch sizes between 50 and 100 kg/day, depending on barrage size and location (Ngoichansri and Thongpun 2003). Annual operating costs are in the range US$176–2,352, mostly for bamboo, ropes, nets and salt for processing fish, and the auction cost varies from US$88 to 8,823. The barrage fishery yields an average rate of return of 150% on total investment costs (i.e., operating plus auction costs).

As a result of both administrative change and the evolution of the rural economy since the 1950s, major changes have occurred in the barrage fishery in LSRB. Formerly, barrage fishing grounds were owned by individuals as a private property. But from 1986 this fishery was reclaimed by communities, and converted to a common property. The reclamation idea was agreed to in 1986 by the community leaders. They wanted the barrage fishery changed from an individually-owned private property, to a common property managed by communities, because (1) income from barrages was required to supplement limited official budgets for community development, and (2) barrage fishing grounds are part of a community's territory, so the entire community should benefit from the income generated, and not just individual and mostly non-resident rights-holders.

However, full implementation of the leaders' decision required 12 years (1987–1999). First, cancelation of individual rights was agreed in 1986, and it was further agreed that from 1987 to 1995 operation of barrages would alternate between original rights owners and communities, after which the right would be held by the communities alone. But implementation during 1987–1995 was difficult. Although the original right-holders lost their benefits as a result of the agreement, they continued to regard barrage fishing as their heritage. Consequently, conflicts and negotiations continued until 2000, when District Officers entered the negotiations, and arranged an agreement among community members and individual owners. Nowadays, most small and low-yielding barrages grounds are still held by individuals, who donate money to the communities. Large barrages are owned by communities, who manage them through an auction system.

5.2.3 The Barrage Fishery: Local Institutions Governing a Common Property

The process begins in April–May, when the community meets to decide the details of the auction. Next, preparations are made to disseminate auction information, either by official letter or during the monthly community committee meeting. The auction is announced for 5 to 7 days in June. Bidding takes place before September, at either Village Halls or Sub-district Council Offices. Both Village Committees and Or-Bor-Tor act as committees to monitor the bidding, to which DoF staff are invited as observers. After the auction, the highest bidder is announced, and contracts signed between the Village Head of the community in which a barrage is located and winning bidder. Normally, the contract defines the rules of barrage operation and bid price payment.

Village heads then announce the exclusion of non-rights holders from barrage areas for at least one month before the successful bidder begins fishing. That announcement signifies the temporary return of the common property rights (barrage fishing grounds) to a private property rights regime (highest bidder) for about seven months, the exact time depending on the water level during the period contracted, from when the auction ends until the end of barrage fishing period. After fishing has finished, barrage areas return to a common property status, and can then be fished by all community members, whose activities must accord with the Fisheries Law.

The process demonstrates that communities are able to ensure that fishers comply with state law by involving the Or-Bor-Tor and the DoF in the bidding process, because they know that both have authority in natural resources management according to both the Thai Constitution (of 1997 and revised in 2007, but with the sections relating to local management unchanged) and the Fisheries Law of 1947. Although DoF officers are loathe to become involved officially in the process, because the barrage is an illegal fishery, their presence as observers indirectly ensures the auction system. The village committees and Or-Bor-Tor members play different roles; definition of the barrage locations to be auctioned and establishment of minimum bid prices are the responsibilities of the village committees, whereas Or-Bor-Tor members are involved in making the bidding arrangements and allocating income from the auction.

Essentially, the auction system represents the formal collective agreement of the community members to pursue those common interests that have no negative effect on any of them. Their decisions are based on a consideration of three main factors: (1) the budgetary requirements for community development, (2) the locations of barrages to be auctioned (they should be far from the village and have the community members' consent), and (3) the number and minimum bid prices of barrages to be auctioned. This fluctuates depending on regional water levels. Fewer barrages are auctioned in years with below average rainfall and therefore lower than average regional water levels, so as to permit all community members to fish and thereby maintain a supply of food to all households. These decisions are made by the community members at the meeting preceding an auction.

Because of the high costs incurred in operating a barrage, in former times city-based capitalists were usually the highest bidders at auctions, and conversely it was difficult for poor people to participate. To overcome this, in 1997 the auction committee revised the rules on auction payments. Now a winning bidder can pay 50% of the total bid price on signing the contract, and the balance either one month later, for outsiders, or after the fish harvest, for community members. As a result, more community members now are able to make the highest bids, either as individuals or as a partnership of five to seven persons who share the investment and labor inputs.

Since 1987 income from the auctions has been shared among communities, Sub-District Councils and District Offices, although the share proportions have changed during the last 20 years. During the initial period (1987–1992) revenues were shared at a ratio of 40:40:20, respectively. From 1993 to 1996, when the government promoted a decentralization policy, the District Office received no share, and income was divided between the communities and Sub-District Council at a ratio of 60:40. Then, in 1997, the Or-Bor-Tor was established to replace the Sub-District Council Office, and the sharing between village and Or-Bor-Tor was re-set at a ratio of 70:30. In 1999 the share was changed yet again, and since then all income from barrage auctions goes just to the communities.[2]

Nowadays, possession of the de facto rights for the barrage fishery alternates between the community and individuals. Communities collectively agree to auction barrages and decide access and use rules for them. Winning bidders are the authorized users, since they have only operational rights of access and withdrawal, and cannot establish management and exclusion rules. However, they can transfer and sell their harvesting rights, as when they sell them to small-scale fishers, and others may access the barrage areas for collecting wild foodstuffs, but not for fishing. Finally, after barrage operations cease the fishing grounds again become a common property open to the entire community (Table 5.6).

5.2.4 *Conflict Between Local and Legal Rights in Fisheries Management*

Serious problems have occasionally arisen since communities began auctioning the rights for barrage fishing, such that most fishers believe that the system has had more negative than positive impacts. A particular grievance is that the system enables a

[2] The only exception is in Phon Klam Sub-District, Sakhon Nakhon Province, where 10% of the income goes to the Or-Bor-Tor.

Table 5.6 De facto rights of the different categories of rights holders under the barrage auction system during auctioned and non-auctioned seasons

Category of rights-holder	Rights during auctioned season (September–January)	Rights during non-auctioned season (February–August)
Communities	Collective choice rights that regulate use patterns and sale as well as exclusion of non-rights holders	Management by maintaining constant rate of use to fit the legal requirements
Highest bidders	Access and withdraw rights as authorized users; fishing and transfer or sale of harvesting rights allowed	Access and withdrawal rights under both state and customary laws that allow harvest of both fish and other wild foodstuffs
All other residents	Access rights; collecting of wild foodstuffs except fish	Access and withdrawal rights under both state and customary laws that allow harvest of both fish and other wild foodstuffs

few wealthy individuals to exploit fisheries resources commercially and destructively while excluding the many small-scale, subsistence fishers. That inequality of access has led to conflict among fishers and their community representatives; between local communities and local DoF officers; between small- and large-scale fishers; between bidders (rights-holders) and non-rights holders over barrage fishing grounds; and among communities with and without barrage fishing rights.

Although most people agree with the Fisheries Law concerning the illegality of the barrage fishery, nevertheless the fishery is widespread in the LSRB, where it has gained increased political and economic importance under the auction system. This indicates that the local institutions are in conflict with the legally constituted national institutions, and whenever DoF Officers attempt to enforce the law, conflict immediately arises between them and local communities. Consequently barrage fishing is tolerated by government, even though it is known to threaten the sustainability of fisheries resources in the LSRB.

That raises the issue of the sustainability of fisheries under the auction system. Most fishers and Fisheries Officers regard barrage fishing negatively, for both ecological and social reasons. Fishers regard it as the most destructive fishing gear in LSRB, because it harvests juveniles directly, and damages brood stock and fish habitats, which result in a long-term decline of fish stocks. Further, it is socially deleterious because it obstructs other fishing gear and therefore excludes other fishers.

However, barrage fishing produces the highest fish yields of all large-scale gear used in the LSRB. Since this is important to the local communities' objective of maximizing revenue, rules are relaxed when applied to barrage fishing, and local DoF officers do not monitor compliance. As a result, barrage fisheries are operated with no regard for long-term sustainability. Further, although the auction system is based on collective agreement and the principle of equally shared benefits, because barrage management is confined to just a local community it does not take into consideration the need for sustainable management of fish stocks throughout the entire Songkhram Basin. Moreover, most fishers do not agree on the barrage auction system, because (1) barrages are the main cause of fisheries degradation; (2) the auction system excludes small-scale fishers, who depend heavily on the fisheries resources; and (3) auction income is no longer important for community development, which, following decentralization, is now funded by the Or-Bor-Tor, and because the income from auctions has been declining in tandem with the decline in fisheries resources.

5.2.5 Degree of Traditional Collective Action and Decision Making

To ensure the realistic planning of any intervention in resources management, it is necessary to understand the knowledge of fishers on management measures. Table 5.7 shows the knowledge fishers possess regarding the establishment process and current management measures of fisheries. Generally, they know about the fisheries management measures within LSRB. The weighted average index (WAI) is at a high level for both groups of fishers.[3] However, there are significant differences in the index of fishers' knowledge. The establishment of fisheries management measures was understood by most respondents, which shows that information about them has been transmitted well. This is understandable because the regulations for fisheries management within the LSRB have been introduced since 1997 via intensive capacity-building programs.

However, there is a significant difference between large- and small-scale fishers, which shows they have different points of view on this issue. Both groups understand that fisheries management has both formal and informal components. The statement 'fisheries management should be implemented for the whole community' is responded to differently by large- and small-scale fishers. Some large-scale fishers do not agree with that statement, which reflects that they ask for different

[3]A weighted average index (WAI), used to analyze the perception and attitudes of the fishers toward fisheries concession management, was computed by:

$$WAI = \sum (fi * Wi) / \sum fi$$

where WAI = Weighted Average Index of Attitude; Fi = Frequency, and Wi = Weighted. Four groups of performance indicators were used to evaluate the outcomes or the efficiency of management as applied in the LSRB: equity, efficiency, sustainability, and degree of participation. A paired comparison of differences between each indicator before 1997 and after 1997 was analyzed to obtain the fisher's perceptions toward changes in performance indicators before and after the decentralized management system was implemented for the whole country. A ladder-like scale was used to measure the perception of fishers, by asking them to make an ordinal judgment of the situation with little demand on their memory, and which could be rapidly administered. The ladder-like diagram used comprises 10 scores, where 10 represents the best situation and 1 represents the worst of their perceived condition of indicators in various time periods. The non-parametric t-test was used to analyze the significance of change in performance indication of management between these two different times.

rules for different parts of country, depending on local conditions. Most small-scale fishers agree that the rule should be implemented for the whole country. This different perception happens because large-scale fishers need special regulation, particularly for their business activities. Some fishing tools used by large-scale fishers are illegal. If the rules are implemented for the whole country, they would become worse off.

However, both groups share the same point of view regarding the statement 'the community rules cannot be enforced for other communities', because they know that different communities have specific rules especially for fisheries resources management. The perception of the two regarding public hearings as the channel for local people to participate in decision making is significantly different, although still within the high level of WAI. Large-scale fishers show weaker support for that statement than do small-scale fishers, with the former considering that a public hearing alone would not enable them to participate in decision making. It means large-scale fishers need more authority in decision making, whereas small-scale fishers felt a public hearing is enough for them to participate efficiently. The response of fishers to the statement 'fishing rule making is responsibility of government only' is not significantly different, but the level of WAI differs between the two, with the large-scale fishers' response being moderate and the small-scale fishers' response in the high level. The perception demonstrates that the large-scale fishers would like to participate more in decision making than would small-scale operators.

Generally, knowledge of both large- and small-scale fishers regarding the current fisheries management measures is at a high level of WAI. However, for some aspects large-scale fishers have less knowledge than do small-scale fishers (Table 5.7). Only four out of 19 statements reveal a statistically significant difference between large- and small-scale fishers. First is that the 'provincial governor has authority in notification of fishing grounds'. That is responded to differently, with large-scale fishers being unlikely to support it whereas small-scale fishers are likely to. Second, the statement 'using bag a net to catch fish is illegal' is responded to differently by the two groups. Although the statement is acknowledged by the large-scale fishers, their support is less than the small-scale fishers. Third, the statement 'installation of brush parks as refuges for brood stock is a fisheries management measure' is also responded to differently by the two groups, with the large-scale fishers responding less positively than the small-scale. Fourth is the fishers' response to the statement 'the harvest of fish in private ponds and release of fish for breeding the next year is a fisheries management measure' is also significantly different. Small-scale fishers are likely to support this measure, in order to maintain the fish population, whereas large-scale fishers give it less support.

Participation of fishers from different groups and different locations in village development, including fishery management, is moderate (Table 5.8). Of all fishers, 50.9% participate moderately in decision making, whereas only 37.1% participate at a high level. Small-scale fishers participate in decision making at a higher level than do large-scale fishers, and those from middle part of LSRB tend to participate more actively than those from other locations. This is because the middle part of LSRB has the highest level of infrastructure development and more social facilities,

Table 5.7 Knowledge of fishers on fisheries management

	Large-scale fishers		Small-scale fishers					Level of
Knowledge of fishers	WAI	LK	WAI	LK	t-test	Sig	Total	knowledge
Knowledge on establishment process of fisheries management measures								
1. Fisheries management measures have both formal and informal types	0.95	HL	0.96	HL	−0.37	0.71	0.96	HL
2. Fisheries management should be implemented for the whole country	0.59	ML	0.78	HL	−3.11*	0.00	0.74	HL
3. Community rules cannot be enforced for other communities	0.76	HL	0.79	HL	−0.56	0.57	0.79	HL
4. The Thai constitution gives a legal right to people in decision making for rules through the public hearing	0.91	HL	0.97	HL	−2.07*	0.04	0.96	HL
5. Fishiery rules making is the responsibility of government only	0.58	ML	0.63	HL	0.65	0.52	0.62	ML
Total	0.76	HL	0.83	HL	−2.50*	0.01	0.81	HL
Knowledge on the current fisheries management measures								
1. Creating an area in front of temples as a conservation zones is a fisheries management measure	0.97	HL	0.99	HL	−1.07	0.28	0.98	HL
2. People believe that some places are haunted, so they do not fish in such areas, and this is a fisheries management measure	0.91	HL	0.97	HL	−1.84	0.07	0.96	HL
3. Provincial governor has authority over notification of fishing ground	0.59	ML	0.78	HL	−3.11*	0.00	0.74	HL
4. Drain ingout of water for fishing is not illegal	0.36	ML	0.48	ML	−1.63	0.10	0.46	ML
5. Using gear to block rivers to catch fish is illegal	0.88	HL	0.92	HL	−1.07	0.29	0.91	HL
6. Using electricity and poison to fish in public water bodies is illegal	0.93	HL	0.95	HL	−0.58	0.56	0.95	HL
7. Using a bag net for fishing is illegal	0.95	HL	0.99	HL	−2.20*	0.03	0.98	HL
8. Prohibition of fishing in a flood area which is spawning ground is a fisheries management measure	0.95	HL	0.92	HL	0.75	0.45	0.93	HL

(continued)

Table 5.7 (conitnued)

Knowledge of fishers	Large-scale fishers		Small-scale fishers		t-test	Sig	Total	Level of knowledge
	WAI	LK	WAI	LK				
9. The spawning season for inland fisheries in Thailand is from 16 May – 15 September	0.91	HL	0.93	HL	−0.49	0.63	0.93	HL
10. Some traditional gear can be used during the spawning season	0.93	HL	0.97	HL	−1.53	0.13	0.96	HL
11. Installation of brush park as the refuge for broodstock is a fisheries management measure	0.88	HL	0.98	HL	−3.34*	0.00	0.96	HL
12. Release of fish on New Year's Day is a fisheries management measure	0.88	HL	0.94	HL	−1.64	0.10	0.93	HL
13. Maintaining water quality is a fisheries management measure	0.97	HL	0.97	HL	−0.11	0.91	0.97	HL
14. Fisheries regulation has been establish for protect fish production for the people	0.98	HL	0.99	HL	−0.21	0.83	0.99	HL
15. Catching only large fish and releasing small ones is a fisheries management measure	0.95	HL	0.94	HL	0.20	0.84	0.94	HL
16. Maintaining an equal number of carnivorous and herbivorous fish is a fisheries management measure	0.90	HL	0.88	HL	0.29	0.77	0.89	HL
17. Harvest fish in private ponds and releasing them for breeding in the next year is a fisheries management measure	0.86	HL	0.97	HL	−3.25*	0.00	0.95	HL
18. Only the Fisheries Law has penalties	0.66	HL	0.75	HL	−1.49	0.14	0.63	ML
19. The best fisheries management is those local people realized and comply with	1.00	HL	0.99	HL	0.89	0.38	0.99	HL
Total	0.87	HL	0.91	HL	−3.45*	0.00	0.99	HL

Note: Level of knowledge (LK); 0.10–0.33 = low level (LL); 0.34–0.66 = moderate level (ML); 0.67–1.0 = high level (HL).

Table 5.8 Participation of fishers in village development, including in fishery management, by fisher category and location

	Fisher category			Location in the LSRB			
Participation	Large-scale	Small-scale	Total	Upper	Middle	Lower	Total
Participation in conflict resolution	100	100	100	100	100	100	100
Very high	8.6	9.9	9.6	6.3	15.7	–	9.6
High	32.8	38.3	37.1	43.8	34.3	35.0	37.1
Middle	56.9	50.9	52.1	50.0	47.9	65.0	52.1
Low	1.7	0.9	1.1	–	2.1	–	1.1
Very low							
Participation in decision making on community development	100	100	100	100	100	100	100
Very high	17.2	11.7	12.9	13.8	17.9	–	12.9
High	31.0	29.3	29.6	26.3	33.6	25.0	29.6
Middle	51.7	57.2	56.1	57.5	47.1	75.0	56.1
Low	–	1.8	1.4	2.5	1.4	–	1.4
Very low	–	–	–	–	–	–	–
Participation in compliance with community rules and decision making	100	100	100	100	100	100	100
Very high	8.6	9.9	9.6	10.0	12.9	1.7	9.6
High	43.1	32.4	34.6	35.0	37.9	26.7	34.6
Middle	48.3	52.3	51.4	52.5	44.3	66.7	51.4
Low	–	4.1	3.2	1.3	3.6	5.0	3.2
Very low	–	1.4	1.1	1.3	1.4	–	1.1

like schools, than other parts. The middle area is more focused on fishing, and has the best developed marketing infrastructure, so fishers from other parts of the LSRB go there to transact business. Participation in decision making regarding community development is also greatest in the middle area, involving around 56.1% of fishers. Large-scale fishers participate more actively than do small-scale fishers (Table 5.8).

5.3 Fishers' Perception of Collective Action and Responsibility for Fisheries Management

The collective action of community members is important for the management of fisheries resources, especially in the development of rules. Most fishers (74.3%) claim that decision making for the establishment of rules related to community fisheries is done through either the village meeting or public hearing, by the collective agreement of most community members. Most community members realize that their participation is important, consequently, most aver that the state and the community should share responsibility equally for fisheries management. Although, about 21.1% of fishers mention that the roles and responsibility of local communities in fisheries management must be greater than government, more than half the respondents reported that the responsible persons for fisheries management in their village are selected mainly by the village head (Table 5.9).

Table 5.9 Perception of fishers toward popular participation and the responsible person in fisheries management

Importance of local participation in fisheries management	Frequency	Percentage (%)
1. WHO MAKE DECISIONS TO ESTABLISH RULES RELATED TO FISHERIES IN THE COMMUNITIES	280	100
• Only formal community leaders and community committees (official)	6	2.1
• Informal community leaders (elders, interested groups within the village)	29	10.4
• Joint decision making between formal and informal community leaders	37	13.2
• Decision making through the village meeting/public hearing process	208	74.3
2. ATTITUDES OF LOCAL PEOPLE TOWARD IMPORTANCE OF PEOPLE'S PARTICIPATION IN FISHERIES MANAGEMENT	280	100
• Very important	48	17.1
• Important	114	40.7
• Neutral	111	39.6
• Unimportant	7	2.5

(continued)

Table 5.9 (continued)

Importance of local participation in fisheries management	Frequency	Percentage (%)
3. WHO SHOULD BE RESPONSIBLE N FOR LOCAL MANAGEMENT OF FISHERIES L	280	100
• Only state	3	1.1
• State more than communities	19	6.8
• Equal shared responsibility between state and communities	188	67.1
• Communities more than state	59	21.1
• Only communities	10	3.6
• Neither state nor communities	1	0.4
4. HOW THE CURRENT RESPONSIBLE PERSON FOR FISHERIES RESOURCES MANAGEMENT OBTAINED RESPONSIBILITY		
• Selected by the village committees	50	17.9
• Selected through the village meeting	50	17.9
• Selected by the village head	151	53.9
• No responsible person in the village – it is the responsibility of the village head	29	10.4

5.3.1 Attitudes of Fishers Toward Leadership

All fishers' thought their community leaders should be accredited for community development as well as fisheries management, and most mentioned that they should be acceptable to and respected by most community members. A number of persons were suggested as those most respected by the community and as being highly influential in promoting sustainable fisheries management; the formal village leaders including village heads, Or-Bor-Tor members and informal leaders who are older, local teachers, and monks. However, the most respected person is the village head, as mentioned by 63.9% of fishers. About 19.4% claimed that Or-Bor-Tor members are respected by the community. In addition, fishers respected and were satisfied at middle and high levels with the effectiveness of the present village leaders.

5.3.2 Customary Rules: Community Management of Fisheries Resources

5.3.2.1 Local Communities Establish Local Fishery Rules to Manage Community Ponds

Many communities in the LSRB manage natural or manmade ponds within their boundaries. They are not auctioned, but reserved usually as fishing grounds for both the community and outsiders. Local rules were established for them via the collective choice of community members for controlling the use of pond water and animals. Fishing practices should comply with community rules, and normally the most destructive fishing gears, such as electro-fishing, poisoning and surrounding nets, are forbidden. However, fishing varies among the communities. Some have established rules that reserve specific use rights only for community members to harvest, for example, food for community ceremonies. Harvesting rights may also be given to individuals, which still requires the agreement of the community committee. For example, the poor are permitted to harvest fish for use at funerals. However, use rules regarding location, amounts harvested and allowed gear are defined by village committees.

The other widespread traditional practice in community pond management is a seasonal restriction on fishing, normally from June to April. Each year the village leaders, community elders, or guardians of a particular pond announce a day, based on Animist traditions and the lunar calendar, when all villagers and people from neighboring villages are allowed to harvest fish communally from the community ponds. After that everybody is allowed to fish the pond until it dries out. Implementation of community pond management varies significantly among villages. In some absolutely no harvesting is allowed before the designated day, whereas in others limited harvesting is permitted. In many cases ponds previously managed as common property have become open access. Those near Buddhist temples are sometimes protected by monks, who, before the ponds completely dry out at the height of the dry season, encourage villagers to rescue fish from them and return them alive to the Mekong River.

5.3.2.2 Local Community Establishment of Fish Conservation Zones

The establishment of fish conservation zones (FCZs) is an important strategy in the LSRB to enhance fisheries production and increase awareness of resource conservation. The FCZs in the LSRB are basically either year-round or seasonal, with fishing normally allowed only during the June–October flood period. There is a total of 16 FCZs in 11 villages ranging in size from 20 ha to about 0.00032 ha, and with a mean size of 9.88 ha. People realize that FCZs increase fish stocks and that fish catches have also reportedly increased after FCZs have been established. Generally it is believed that the impact of harvesting can be reduced by banning or significantly limiting fishing activities in key deep-water areas that serve as dry season refuges and sometimes spawning grounds. Community rules to control and manage FCZs were established through community meetings, and committees are also assigned to monitor FCZ management. Most fishers accept that popular participation in decision regarding rule-making is very important. Most agree with the community rules for management of FCZs; about 81% of them agreed fully. The high level of agreement was reached because the rules for FCZ management are established by the community members. The establishment of an FCZ is recognized as the most effective strategy with a high degree of participation between government and local people, because it has more positive than negative impacts.

5.4 Conclusion

Fisheries resources in the LSRB are managed under a complex multiple-rights property regime, by which individual, common and state property rights are defined and both combined and separated. This has resulted in overlap, conflict and complementarities, and in a varied performance (Khumsri et al. n.d.). Changes in external economic and political contexts led to change in LSRB property rights, via a lengthy process characterized by struggle and negotiation, as both original individual rights holders and communities adjusted to evolving institutional arrangements. In many instances, however, problems within communities were resolved by coercion from external government. Further, as was demonstrated by the example of the barrage fishery, changes in property rights regimes may be multi-directional.

Institutional arrangements also change concurrently with property regimes, owing to structural changes in rights and duties that link people and resource systems. This was demonstrated when LSRB communities established new fisheries management institutions, by combining national institutions with village committees and Or-Bor-Tor. Moreover, communities also respect multiple types of property rights allocated locally to both individuals and communities. In other words, they neither rely on one particular kind of property rights regime or clearly distinguish among the types of property right. This provides incentives to participate in fisheries management through collective action.

On the other hand, without specific rules that situation does not guarantee sustainable fisheries management, as demonstrated by the barrage auction, when the communities' desire to maximize income in turn drives winning bidders to seek maximum profit from the fishery during their very brief exclusive tenure. Without rules aimed specifically at sustainable use, overexploitation and the eventual collapse of the fishery is inevitable.

Rights to manage fisheries have varied sources and are exercised differently. Although the Thai Constitution supports natural resources management by communities, there are neither guidelines for practical implementation nor clearly defined authority and roles. For example, national law may grant LSRB communities de jure rights of access and withdrawal, while reserving for government the formal rights of management, exclusion and alienation. Yet concurrently the communities hold de facto rights to manage fisheries within their boundaries. Thus there is duplication and a mismatch between local and state institutional arrangements for fisheries management.

Acknowledgements All field research and related activities were supported financially entirely by the Fisheries Program of the Mekong River Commission, to which I am most grateful. I am especially grateful the many people in Lower Songkhram River Basin who cheerfully tolerated my intrusions and questions, and unstintingly shared their experience and knowledge.

References

Blake, D. J. H. (2006, Sept). *The Songkhram River wetlands – a critical floodplain ecosystem of the lower Mekong Basin*. Paper presented at the International River Symposium 06, Brisbane.

Boonyaratpalin, M., Kohanantakul, K., Sricharoendham, B., Chittapalapong, T., Termvidchakorn, A., Thongpun, W., Kakkaew, M. (2002). *Ecology fish biology and fisheries in the Lower Songkhram River Basin* (Technical paper No 6/2002). Bangkok: Department of Fisheries (in Thai).

DEDP (Department of Energy Development and Promotion). (1997). *Revised feasibility study on the Songkhram project. Main Report*. Bangkok: Department of Energy Development and Promotion (in Thai).

DEQP (Department of Environmental Quality Promotion). (2002). *Seasonally flooded-forests*. Bangkok: Ministry of Science, Technology and the Environment (in Thai).

Hortle, K. G., & Suntornratana, U. (2008). *Socio-economics of the fisheries of the lower Songkhram River Basin, Northeast Thailand*. (MRC Technical Paper No. 17). Vientiane: Mekong River Commission.

Khumsri, M. (2008). Co-management institutions for sustainable inland fisheries management: A case study of the lower Songkhram River Basin, Thailand. Dissertation, Asian Institute of Technology.

Khumsri, M., Ruddle, K., & Shivakoti, G. P. (2009). Rights and conflicts in the management of fisheries in the Lower Songkhram River Basin, Northeast Thailand. *Environmental Management, 43*, 557–570.

Khumsri, M., Ruddle, K., Sriputinibondh, N. (n.d.). Decentralized fisheries management: Effectiveness of the Or-Bor-Tor in managing the inland fisheries of the Lower Songkhram River Basin, Thailand. In Anon. (Ed.), *7th Technical symposium on Mekong fisheries, MRC conference series*. Vientiane: Mekong River Commission (in press).

KKU (Khon Kaen University). (1997). *Study on protection planning for the impact of the Nam Songkhram Project on the environment*. Khon Kaen: Khon Kaen University (in Thai).

MRC (Mekong River Commission). (2000). *Preliminary results of a fisheries survey of the Songkhram River basin, N.E. Thailand*. Udonthani: Mekong River Commission (in Thai).

Ngoichansri, S. & Thongpun, W. (2003). *Fish production of the barrage fishery in the Lower Songkhram River Basin*. Annual Report of Sakhon Nakhon Inland Fisheries Development Center (pp. 58–91). Sakhon Nakhon: Department of Fisheries (in Thai).

OEPP (Office of Environmental Policy and Planning). (1999). *The wetlands of the Northeast*. Bangkok: Ministry of Science and Environmental Technology (in Thai).

Ostrom, E., & Schlager, E. (1996). The formation of property rights. In S. Hanna, C. Folke, & K. G. Mäler (Eds.), *Rights to nature: Ecological, economic, cultural, and political principles of institutions for the environment* (pp. 127–156). Washington, DC: Island Press.

Rainboth, W. J. (1996). *FAO species identification field guide for fishery purposes, fishes of the Cambodian Mekong*. Rome: Food and Agriculture Organization.

RTG (Royal Thai Government). (1930). *Thai civil and commercial law of 1925*. Bangkok: Royal Thai Government (in Thai).

Ruddle, K. (1994). External forces and change in traditional community-based fishery management systems in the Asia-Pacific Region. *Maritime Anthropological Studies, 6*(1–2), 1–37.

Sricharoendham, B., Dongsawat, S., Chubchavang, B., & Rattachumnong, D. (1998). *Fish diversity and rehabilitation of fish communities in the Pong Chi Mun River*. Bangkok: Department of Fisheries (In Thai).

Yingcharoen, T., & Vilapat, C. (2000). *Survey of ecology and fisheries in the Songkhram River in Nakhon Phanom Province*. (Technical Paper). Bangkok: Department of Fisheries (in Thai).

Chapter 6
Vietnam: The *van chai* System of Social Organization and Fisheries Community Management

Duy Thieu Nguyen and Kenneth Ruddle

Abstract The historical evolution of *van chai* from within the administrative structure of farming villages is explained, and their geographical distribution described. The social and management functions and administration of the 'floating village' type of van chai is explained with particular reference to the lagoons of Thua Thien Hue Province. The van chai is the focus for the spiritual activities of fishing, so in each fishing community founded by migrants along the South-Central coast a van chai was established to worship the Whale God. These reflected the traditional folk and professional beliefs and mutual assistance within the community. Analysis of the religious and social functions and organization of the 'guild-type' van chai of the South-Central coast is based on Van Thuy Tu, Phan Thiet City, Binh Thuan Province. The general design principles of pre-existing management systems in Vietnam are examined in terms of rights, rules, monitoring and accountability, conflict resolution, and sanctions.

Keywords Customary law • Floating village • Local religion • Village administration • Whale shrine

---

D.T. Nguyen
Vice Director and Associate Professor, Vietnam Museum of Ethnology,
Nguyen Van Huyen Road, Hanoi, Vietnam
e-mail: thieund_dna@yahoo.com

K. Ruddle (✉)
Asahigaoka-cho 7-22-511, Ashiya City, Hyogo-ken 659-0012, Japan
e-mail: mb5k-rddl@asahi-net.or.jp

K. Ruddle and A. Satria (eds.), *Managing Coastal and Inland Waters: Pre-existing Aquatic Management Systems in Southeast Asia*, DOI 10.1007/978-90-481-9555-8_6,

6.1 Introduction

In many parts of Vietnam there is a long tradition of local fisheries management and mutual assistance, operating through institutions known as van chai.[1] They have been important in fisheries administration, principally in managing the numbers of fishers, tax collection, social relationships, and harmonizing fishing operations. The van chai is a comprehensive institution structured to address the basic issues of community and aquatic resources management. These are (1) shrine management and the conduct of ceremonies; (2) mutual assistance among fishers; (3) specification of the behavior, rights and obligations of fishing boat owners, captains and crewmembers; (4) disposal of the catch; (5) governance of fishing operations; (6) specification of the rules for the main gear types (pertaining mainly to eligibility, seasonality and profit-sharing); (7) conciliation of fisheries conflicts, the resolution of which is not stipulated in current local rules or higher laws; and (8) sanctions (punishment) (Ruddle 1998). Although details vary considerably by locality, the underlying principles of the veneration of deities and ancestors combined with the sacred obligations of mutual assistance remain all pervasive, and underpin all other objectives by providing the van chai with its moral authority (Fig. 6.1).

Basically, however, the Vietnamese have been always been farmers, and historically the intensity and type of fishing activities varied considerably within the country (Nguyen D.T. 2009). Inland fishing activities in the lowlands of the north and south never were of major importance. Further, the northerners were unfamiliar with seafood, such that many spices were used during cooking to mask the taste of marine fish and make them resemble freshwater species. Southerners started fishing much later than did the northerners. In addition, since the south of Vietnam is rich in freshwater fish, southerners never traditionally sought marine species (Nguyen D.T. 2009).

The situation is different in the Central Region, where agricultural land is poor and scarce, and the swift-flowing rivers sustain few fish. There, in contrast to the northern and southern parts of the country, marine currents bring large fish stocks into nearshore waters. So migrants from the northern provinces of Vietnam who settled in coastal area of the Central Region became marine fishers, and a new culture and way of life gradually emerged.

Because van chai emerged from farming villages and gradually engaged in community self-management, their administration and social management reflected traditional Vietnamese agrarian culture. Further, a strong emphasis on the local community resulted in a diversity of cultural characteristics among the van chai of different localities.

[1] The term *van chai* means a community that lives on its boats and earns a living on rivers or in coastal areas by fishing with simple, small-scale gear. The term '*chai*' means 'gear used by small-scale and artisanal fishermen' (Anon 1988).

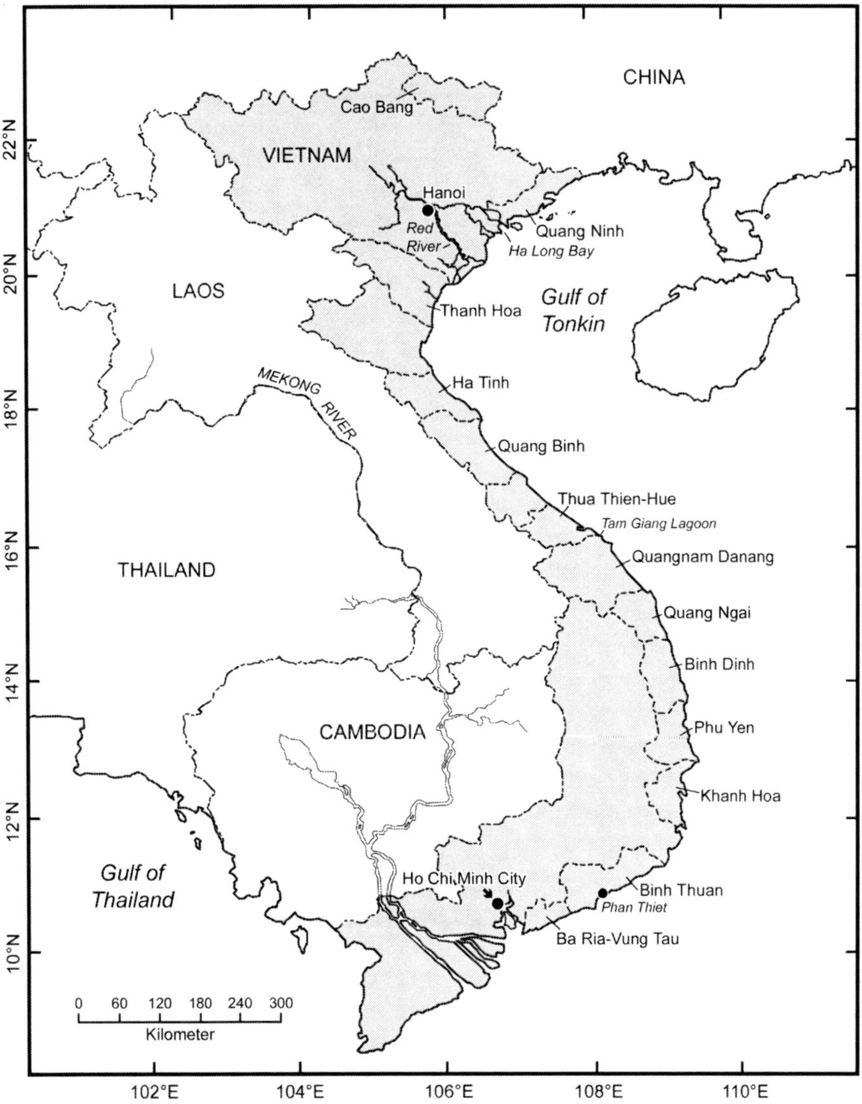

Fig. 6.1 Locations referred to in Vietnam

It is important to understand that the term '*van'* has two meanings (Ha and Nguyen 2009). One is an organization of persons who follow the same profession. In this sense it approximates the English term 'guild'. This is the meaning used particularly in the Central Region. However, as used among riverine fishers the term '*van'* means a 'village'; i.e., in that case *van* is an administrative unit. In earlier times the village was the common agriculture administrative unit of Vietnam, and was defined as a rural population forming a specific settlement unit that comprised the lowest level administrative unit during the feudal era (Anon 1988). In its sense as an administrative unit the term '*van*' also embraces the so-called 'floating people', who reside on their boats. So '*van*' remains widely used to mean 'floating village'. In this usage of the term the population of a *van* includes all who live together on a section of a river; fishers, various river boatmen, small-scale transportation workers, and small traders. Thus the term '*van*' means a general community of riverine residents. However, it is not used to define people who live on land and earn their living on rivers. Such a community is called a 'fishing village' (*lang ca*), not a *van*. Two types of van chai can be distinguished based on those definitions: (1) a floating village, and (2) a land-based fishing community.

The floating village (*van chai thuy cu*) is group of fishers and their families who live permanently on their fishing boats, who lack both a dwelling house on land and farmland, and who always operate and make a living on rivers or in estuaries or coastal lagoons. This type of van chai exists along the coast from the North to the North-Central regions. Examples are the van chai at Cua Sot, along the Da and Red Rivers, those in Ha Long Bay, and those at Dam Thuy Dien, in Tam Giang Lagoon, Thua Thien Hue Province. The first van chai anywhere in Vietnam were established for inland fisheries, mainly along the lower and middle reaches of rivers in the north, and around coastal lagoons in the north-central part of the country. In comparison, those in marine embayments, like Ha Long Bay, or in enclosed coastal areas, like Tam Giang Lagoon, are relatively recent, the latter having been established about four centuries ago.

The land-based fishing community (*lang ca*) comprises a group of fishers who earn their living by fishing, but who have a house on land, either along a beach or an estuary, or on an island. Some fishing families also cultivate around their houses and raise livestock, to provide for their own subsistence. This type originated from farming populations who during the 'feudal era' participated in officially organized migration to the central and southern areas of Vietnam. Vietnam has abundant and varied inland waters and wetlands, such that locations for fishing are almost ubiquitous. Most farmers also use a variety of aquatic resources for family consumption, livestock feed and sale, and, just like fishers, many farming families living around inland water bodies and in coastal areas make a living by capture fishing. Indeed, so dependent are they on aquatic resources that they are really full-time fishers, and not farmers. Yet they live among farmers and belong to farming villages, the customs and traditions of which they must follow.

In the South-Central Region, the structure of a van chai is not that of a village, but is like a 'guild'. There the van chai is the center of community religious beliefs, and therefore of fisheries and community management.

The structure of this chapter is based on those two types of van chai, since they reflect distinct histories. Further, the nature of the van chai is also different (Ha and Nguyen 2009). The difference emerged because the van chai in the north are ancient, whereas fishing villages first emerged in the south only in the seventeenth century. At that time the social management system of the Nguyen rulers was loose, and directed toward farming villages. Although *van chai* were established by fishing communities, nevertheless they played a key role in enabling the rulers to stabilize coastal communities, and use them as bases for further southward expansion.

In coastal areas the establishment of van chai might have occurred in two stages. In the first a group of fishers, possibly having migrated from the same area, formed a van chai. The first-comers, often the elders and the experienced, were selected to lead this group. A leader of a van chai and his management team would then handle all livelihood and cultural activities. Assisting the leader were a *huong van* (secretary for clerical work), *chanh bai* (the person in charge of ceremonies) and *thu tu* (keeper of the temple/pagoda). In this stage the van chai became the nucleus of administrative and customary activities of the community (Nguyen 2002a). During the second stage the Nguyen rulers became stronger and more capable of controlling the South-Central Coast, and gradually they established their three-tier government of hamlet, district and commune. The administrative power of the State then reached the fishers' van chai, and from that time it would have been under the control of the lowest administrative level of the State. The van chai was understood basically as the focus of spiritual activities of fishers residing in a particular area (Nguyen 2002a).

6.1.1 Village Structure and Management: A Prerequisite to Understanding the Van Chai

The village community has long been the basic administrative unit in Vietnam. Since van chai originated from farming villages, it is necessary to digress somewhat to explain the village as the basic unit of society.

It has long been assumed that the Viet village was relatively independent of central authority, existing as an autonomous, closed and independent small territory within a larger area. That, however, is not entirely accurate, since from the tenth century, as Vietnam gradually became independent of China, attempts by the central government to impose its will on the village were frustrated by protracted civil war (Ha and Nguyen 2009).

Farming villages managed themselves under the 'five-notables or five chiefs' system, which comprised a mayor, deputy mayor, village notable, secretary, and head night-watchman. The village mayor coordinated the general work of the village, in which he was assisted by the deputy village mayor. The village secretary assisted the council of the 'five-chiefs'. The village notable took care of such public construction as roads, the communal house, the temple, and the pagoda. And the head night-watchman was responsible for village security. Each village had a series of subordinate hamlets with hamlet heads. Besides implementing several responsibilities to the State, such as supervising and expediting tax collection, corvée labor and military conscription, the village managers dealt mainly with such internal affairs as security, funerals, weddings, and, in particular, worship (Ha and Nguyen 2009).

However, centralization efforts were eventually revived. The Later Le Regime (1428–1788) focused on centralizing the national administrative system, and in 1428, King Le Thai To subdivided Vietnam into a hierarchy of micro-, meso- and macro-communes. Villages were ruled by a communal or village mandarin, who became the village chief during the reign (1460–1497) of Emperor Le Thanh Tong. The communal or village mandarin was subordinate to the district and *chau*.[2] In addition to the village chief or mandarin, there was a 'Council of Village Notables'. comprised of the wealthy elite (Ha and Nguyen 2009).

The reforms made during the regimes of the first Le kings marked a turning point in natural resource control and management by the central government during the feudal era. Among the most important issues were policies for state or public land. Public lands were reallocated to villagers every six years, based on their social status. Allocated lands could neither be sold nor inherited. The main objective under the Le rulers was to put public lands under the control of the Central government. And they were relatively successful in separating villages from their resources, which then became directly controlled by the central government.

However, from the middle of the sixteenth century civil wars again absorbed the attention of leadership more than did administration. Consequently, villages became increasingly autonomous and self-reliant as the rural population sought to ensure its own survival. By the end of the Le Era village autonomy had so revived that the commune remained the only administrative unit that the central government was capable of managing.

To facilitate their rule, the French Colonial Administration (1895–1954) retained the pre-existing organization. However, they separated the country into three regions, each administered differently.[3] As a result, although the villages were maintained they were managed and ruled differently in the three regions. Regardless of location, during this period villagers had to obey both their existing indigenous leaders and the French colonialist regime. By decrees issued in 1904 and 1925 the Vietnamese kings were stripped of their supreme ownership of national properties, including coastal waters, which were declared 'national common property', owned by the French government. Where a van chai used large fixed gears, the fishing rights and the rights of the water management belonged to the *van,* and were recognized and protected by State authority. Elsewhere the fish resources of a river or the sea were considered open access and could be freely exploited by anybody. In contrast, pre-existing community-based fisheries management continued virtually unaltered (Nguyen 1995).

A communal authority system was introduced after 1945 in the northern part of Vietnam. As a result, the relationship between the State and village/commune

[2] When under Chinese rule, the term *chau* referred to an administrative unit equivalent to the present-day 'District'. *Chau* remained in use in mountainous areas under the Nguyen Regime, but was discontinued after 1945.

[3] The South (so-called 'Cochin China') was a colony, the Center (so-called 'Annam') a protectorate, and the North (so-called 'Tonkin') a protectorate under direct French rule.

changed, as did the village/commune administrative structure (Ha and Nguyen 2009). From 1945 to 1955, the Council of Village Notables was replaced by an Administrative Resistance Committee (during the anti-French War) and an Administrative Committee.

After 1975 the administrative structure of the now re-unified Vietnam was changed again, with the formation of the Communal People's Committee and the People's Council. The positions of village/commune mayor and deputy village mayor became Chairman and Deputy Chairman of the Communal People's Committee, respectively. This democratically elected committee constitutes the lowest level of governmental authority.

6.1.2 Key Differences Between Water- and Land-based Villages

There are key differences between the water-based hamlets of a 'floating village' van chai, and a group of land-based hamlets linked to a farming village. One is that the latter have a fixed location, whereas a constantly mobile 'floating village' hamlet does not. Another is that whereas the communities of farming villages adhere to the territorial principle, and are linked with each other in a neighboring relationship, communities in the water-based hamlets/villages respect blood and professional relationships. In a water based hamlet people with the same family name always use the same fishing gear and gather together to form a hamlet. Other important differences are that a van chai hamlet lacks the well, public watch-post, Kitchen God temple, and security guard team that characterize farming villages. Therefore, compared with a farming village, in general a 'floating village' type van chai is a community unit that lacks the so-called 'real' administrative spirit and meaning. Rather it is a traditional, small and self-managed community (Nguyen D.T. 2009).

The blood relationship among van chai residents is close. This is demonstrated by boat mooring places in a lagoon, where some five or more family boats are always moored together. These kinship groups constitute the basic administrative unit of the *van*, through their representatives in the van council meeting. In addition to being morally and materially supportive, the blood and family relationships of the van chai constitute an essential locus for the vital intergenerational transfer of technical knowledge and professional secrets, normally kept within the small circle of family members (Ha and Nguyen 2009).

A 'floating village' type a van chai is also characterized by professional type, with, for example, fishers who use a particular gear type forming a van. In Vinh Ha Commune on Tam Giang Lagoon, Thua Thien Hue Province, for example, the 12 van chai are distinguished by gear type (Tran 1996a), although gear types may be switched seasonally. However, sometimes there is occupational diversity within the membership of a van chai. For example, although the members of Ky Xuyen van, located in Sot Port, Ha Tinh Province live together, they use different fishing gears (Nguyen 1984, 1993). Similarly, although most members of a van are fishers, some earn a living by using their boats to ferry passengers or transport freight. The number has been increasing continuously, as fishers and their children change to jobs that are usually both more profitable and easier than fishing. However, job change does not affect the attributes of the van chai, since all members of the community must obey the same rules on boat mooring (see below), as well as continue to perform the traditional rites (Nguyen 2003).

'Floating villages' can also be distinguished by the administrative role played by age class (*giap*) Originally the term '*giap*' referred to a Chinese administrative unit used for making a census (Nguyen 2002b). However, when imported to Vietnam millennia ago the meaning changed; now it means a place where village males of the same age class gather. Its official administrative meaning was discontinued. Despite that ancient change of main meaning, the giap continues to play a particular administrative role in a van chai. Unlike farming villages, members of a 'floating village' type van chai possess neither taxable land nor a fixed residence. Therefore the authorities operated through the giap and family organizations to record information on all village families.[4]

6.2 Floating Village

There have never been many 'floating villages', and they are not widely mentioned in the literature (Ha 1995, 1998). Yet it was observed that early in the twentieth century Vietnam had many 'floating villages', which included groups of fishers or boatman (Dao 1938), although numbers were not provided. Other sources indicate that the number of floating villages was not high. At the beginning of nineteenth century there existed 70 floating units in 12 old towns in the provinces of the Red River Delta and along the coast from Quang Ninh Province to Ha Tinh Province (Hoang 2003:5). In the 1930s, there were about 90 floating hamlets or villages located in rivers, and about 21 along the coast in the region from the Vietnam–China border to the Tien Yen area (Nguyen 1995:24).

In contrast, in the Central Region from Thanh Hoa to Binh Thuan provinces, where rivers are short and the land mountainous, freshwater bodies cannot sustain fisheries large enough to support 'floating villages'. Therefore the only such communities are concentrated in estuaries and lagoons.

Since 1955 the number of 'floating villages' has decreased, mainly as a result of the reorganization of rural management, irrigation development, the destruction of riverine resources, and water pollution. Consequently, many families either switched to living on land or changed occupations entirely.

[4]During the French colonial era information was recorded in a 'Registration Book' issued by the Taxation Agency. It contained the names of boat owners and all family members, their civil status, taxes paid and conscription, among other information.

6.2.1 The Administration of 'Floating Village' Van Chai

Originating from land-based villages, 'floating village' van chai are managed according to a simplified version of the administrative system used in farming villages. It has three main components: (1) males 18 years-of-age and over, (2) The Council of Elders, and (3) village officials. Under the French Administration, on turning 18 years-of-age males had to pay a poll tax, vote and attend village meetings to deal with everyday issues. Although in theory the Council of Elders (local names vary) is elected by the citizens, in practice it includes all the heads of family clans, who represent their members. Only four village officials are required, because public works are fewer and less complex in a van chai compared with a farming village. They undertake the same functions as those in farming villages, as well as performing the self-managed activities for the van, and also perform the administrative functions of the lowest level of the State management system. Unlike farming villages, an important task of the van leader is that of a priest who officiates at funerals and weddings, as well as handling conflict resolution, performing the annual rites and organizing the festivals.

When a van was formed, one person became responsible for general management. After that the head of the van was elected. The van elders submitted a proposal to the authorities asking that their leader be accorded the same rights and obligations as the head of a farming village. In addition to the head, a Council of Notables was also elected. Each family clan in a van nominated a representative to the Council of Notables, which in effect was a Council of Family Heads, because the members were often the head of family clans. Then the van Council decided which of its members would become a chief. When any member of the leadership team either died or resigned, the van had to elect a replacement, and this person had to be approved by the Council of Family Heads. Thus, within an administration system of the van chai, the Council of Family Heads played a key role.

Should a district authority deny a van's proposal to establish its own independent management authority, then it would remain a water-based village belonging to a farming commune. In that case a van's administration system would not be the lowest state administration unit. Although those rules exist, a van has a relationship with a farming commune only in terms of tax payment. There are no other linkages.

6.2.2 Management Structure of Van Chai in the Lagoons of Thua Thien Hue Province

A male leader with occupational experience and prestige headed each van chai. He was the voice of the community, especially in communicating with outsiders. In addition, he was responsible for solving internal disputes by balancing justice, mutual assistance, familial and neighborly relationships, and national standards of ethics.

The leader consulted with the community elders on such vital issues such as migration, site selection for fishing, and for shelter during storms and floods. However, he alone was responsible for making the final decision (Nguyen 2002b).

During the process of electing a leader, candidates were evaluated based on their prestige, sense of ethics and other essential talents. Since feudal governments could not control van chai, such an election was a democratic process. This was apparently unlike the situation in farming villages or communes. The leader was not paid, so like everybody else he had to make living to support his family, while assuming the responsibilities entrusted to him by the community members. The leader often served permanently, although he could be dismissed for a serious breach of ethics.

Since the mid-twentieth century, smaller van chai have tended to merge into larger and more densely populated units. This led to the creation of the post of Assistant to the Leader, who assists in management and public duties, besides functioning as secretary and accountant.

6.2.3 Structure, Relationships and Institutional Formulation in 'Floating Village' Van Chai

The membership principle of a van chai rests on the two key relationships of blood and profession. Thus it is simpler than that of farming villages. The formation of a van chai starts with family relationships and then extends to a group of relatives who gradually intensify their professional and interpersonal relationships to produce a small society with its own specific characteristics (Nguyen 2002b). Relationships with external societies took different forms, and involved neither administration nor dependency.

Van chai have managed fishery activities in lagoons in Thua Thien Hue Province since feudal times. Nguyen Era documents confirm that from the sixteenth to the eighteenth centuries governments attempted unsuccessfully to group fishing communities into administrative units. However, the social structures of 'floating villages' were fluid, which made control by the government difficult (Ha 1995).

In contrast, itinerant livelihoods demand cooperation and mutual assistance among members of fishing communities as well as among 'floating village' van chai, based on an empathy among people who share the same living conditions, and who are often linked by kinship. Because a van chai is a social structure more than an administrative unit, the occupationally mobile fishers can move from one to another, without first completing any administrative procedures. In the lagoons of Thua Thien Hue Province, for example, in-migrants are given rights and obligations similar to other members.

There are several reasons for that. First, the van chai structure is fluid, and migration is very common because its members must remain mobile to make a living. Second, lagoon-residents greatly respect neighborhood relationships, because they live in a natural environment that combines both abundant resources and numerous challenges.

Therefore the need for mutual assistance is great. The struggle for survival has increased the need for mutual assistance among lagoon fishing community members, such that discrimination between in-migrants and earlier inhabitants does not exist. This is a major difference between lagoon fishing communities and farming villages. Further, because lagoon residents were marginalized by feudal society and looked down on by the royal courts, they have tended to aggregate and share egalitarian attitudes, in order to combat discrimination. In addition, unlike land the resource-rich lagoon waters cannot be demarcated for private ownership. Newcomers have no negative impacts on the 'accommodation' and occupation of earlier residents, so no economic dispute arises.

For self-management the authority system of a van chai follows the traditional rules or conventions. Although de facto rules, locally these have de jure attributes. As in farming villages, all such traditional conventions of the van chai are obeyed absolutely. There are three main types of van chai rules and regulations: (1) those pertaining to the place and order of boat mooring (i.e., the residency rules) (Photo 6.1), (2) those covering fishing rights, and (3) those on fishing grounds and resource protection.

Although the fishers of a van chai are highly mobile during their routine activities, they are not nomads. For administrative purposes they are considered as having a fixed residence at the main place where they moor their boat. That address must be registered. When fishing elsewhere they must register temporarily with the local authority. In their home water areas only the *van* members are officially resident, so any outsider must obtain the permission of the local van chai to enter to avoid storms, sell fish, buy supplies or to fish. An outsider refused permission must move on immediately.

Photo 6.1 Residential and fishing boats of a 'floating village' van chai on the Nhu Y River, a tributary of the Huong River, Thua Thien Hue Province, Vietnam

6.2.4 The Relationship between a 'Floating Village' Van Chai and its Host Farming Village

Fishing villages are the largest social community unit on the lagoons of Thua Thien Hue Province. Nevertheless, for administrative purposes their van chai are under farming villages. This difficult relationship, which has bound lagoon residents for centuries, was probably first institutionalized during the Nguyen Era, to control van chai indirectly through farming villages. Because of this, members of a fishing community had to pay taxes and perform military service. Although lagoon residents felt a 'royal pressure' from the host village, they had no major connection to it, other than through the paperwork handled by their leader (Nguyen Q.V.B. 2009).

In addition a van chai was socially subordinate to its host village. The residents of van chai had no land and had to perform the least desirable jobs, so they were scorned as people who 'live with no house and die with no land in which to be buried', and denoted using discriminatory and demeaning terms (Nguyen 2002a). In the spirit of an agricultural and conservative feudal society, with its strictly hierarchical organization, the landless, who had to live on the lands of other villagers, were always classified as servants. For example, during festivals residents of 'floating villages' were forced to provide unremunerated services and were not allowed to sit equally with farmers in the village communal house.

As a van chai gradually became overcrowded with many poor and mobile fishing families, their administration by farming villages became ineffective. Further, because van chai lacked land and their riverine fisheries were open of access, they were of no economic benefit to farming villages. As a result their administrators ignored the van chai. Their members were neither required to perform public service nor assume village responsibilities. In that way most van chai gradually became separated from farming villages, and developed into self-managed units with traditional social characteristics that have remained largely intact until the present.

6.3 The 'Guild-Type' Van Chai

The van chai has been in existence for centuries in the South-Central coastal region as a multifunctional institution, responsible for governance, security, organization of fishing activities, disaster relief, resource conservation, and the conduct of funerals, marriages, and the ceremonies of spiritual life.[5] Fishers' societies in this region are no longer linked

[5]There have been relatively few social and cultural studies on the coast of Vietnam, so the livelihood and other aspects of fishing communities, particularly their formation and the material, spiritual and religious life, is not well understood. Unlike the North-Central Coast and Northern regions, even fewer studies have been conducted on the villages and social structure along the South-Central Coast, so reliable documentation is lacking (Nguyen D.T. 2007, 2009).

with farming activities, as they are in both the north and south of Vietnam (SARI 1998; Nguyen D.T. 1984, 1993, 2002a, b, 2003, 2007, 2009). However, the ancestors of most fishers in the South-Central Region were northern farmers who migrated and then became fishers along the coast of the South-Central Region. During the early days of migration those people who left their families and villages might have returned home to the Northern provinces to participate in social activities with the farmers. Gradually, the new environment of the migrants presumably required new spiritual activities and social forms more relevant to a fishing community. That need to gather for spiritual activities prompted the establishment of van.

6.3.1 The Structure of Traditional Villages in the South-Central Region

The establishment history of villages in the South-Central Region is closely related to territorial expansion of the ethnic Vietnamese toward the southern part of the country, which the Nguyen rulers began in 1693. Initially soldiers of local garrisons were permitted to marry local women and establish villages, but settlers from the North-Central provinces of Quang Nam, Quang Ngai, Binh Dinh, and Phu Yen, were later mobilized to farm more land in the South-Central Region. While communities established a van chai for self-management, the Nguyen rulers applied the *Dao* (province), *Tong* (district) and *Lang* (village) three-level social organization structure of the North, to establish the administrative system in the South-Central Region (Nguyen D.T. 2009).

Confucian scholars selected the person running the Tong or District. He was known as the *Cai Tong*. *Lang* or villages were supervised by the *Ngu Huong* (Five Official Ranks of the Village), namely *Ly truong* (Headman), *Pho ly* (Assistant Headman), *Huong veec* (Secretary), *Huong muc* (Head of Rituals), and *Huong kiem* (Village Security Officer). *Ly truong* assumed general administration and was supported by the *Pho ly*. The *Hương viec* was in charge of clerical work, the *Huong muc* took care of public works such as roads, temple, communal hall and pagoda, and it was the duty of the *Huong kiem* to maintain safety. Those five officials comprised the management team of the village. In addition to State obligations, such as tax collection and conscription for military service or corvée labor, they were responsible for handling all village matters, particularly funerals, marriages, security, and ceremonies in the communal hall.

The general process has been described for Binh Thuan Province, where in the late-seventeenth century the southern borders of what is now Vietnam were around the present-day city of Phan Thiet. To stabilize the frontier in newly conquered territory, the Nguyen rulers brought settlers from the northern provinces of the Ngu Quang Region. The first such settlers gathered in the Phan Thiet area at what is now Duc Thang Ward, where general rural occupations gradually gave way as the marine fishery developed. The van chai started in 1697, and so now has just over 300 years of history as an association (Ruddle 1998; Ruddle and Luong 2009).

Photo 6.2 Main buildings before restoration at Van Thuy Tu, Phan Thiet City, Binh Thuan Province, Vietnam

The Van Thuy Tu shrine was established in the winter of 1762.[6] It initially functioned as a shrine to village deities and ancestors, but as the sea fishery developed that function was displaced by the worship of the Deity of the South Sea. It is also known that various kings approved the fishing rights of Van Thuy Tu in 1843 and 1887.[7] A hamlet in Duc Thang Ward, the first fishing settlement in the region, is regarded as the center from which the whale cult disseminated in Binh Thuan Province (Photo 6.2).[8]

Generally, the Pre-colonial or Feudal Era saw the foundation and spatial broadening of van chai. In Binh Thuan Province, for example, some authors (Republic of Vietnam 1963; and Ruddle and Luong 2009) confirm that the historical development of van chai began at the end of seventeenth-century, when migrant fishers from the central regions settled on the unutilized land of Phan Thiet. Fishers first congregated in small clusters, which then became hamlets (*xom*), van and villages (*lang*). Their livelihoods depended mainly on capture fisheries and fish processing. Under the Nguyen rulers, especially during the reigns of the kings from Minh Mang to Tu Duc (1820–1883), royal settlement policies encouraged more fishers from the north to settle in Binh Thuan Province.

[6] This is known from Chinese characters written on the beam in the main hall of the shrine and verifiable from historical documents and other objects kept in the shrine (Truong n.d.; Anon 1996).

[7] This is known from the original documents kept in the shrine (Anon 1996).

[8] In the main hall of the shrine the originator (sage) of the fishery at Duc Thang Ward village is worshipped as the originator of all fishing villages in Binh Thuan Province (Ruddle 1998).

6.4 Van Chai: The Focus for Fishers' Spiritual Activities

Concurrently with the establishment and development of fishing communities along the coast, each in-migrant community cluster established a van chai to worship the Whale God (*Ca Voi*). These reflected the traditional folk and professional beliefs of their home regions, and resulted in the intensification of mutual respect and assistance within the fishing community. Dozens of temples (*dinh van*) in van chai were established, mostly in the eighteenth and nineteenth centuries, and were closely connected with the formation of communities and foundation of fishing villages (Photo 6.3). The van chai was the first socio-occupational organization in a new settlement, having been established ahead of village and commune governments (Nguyen D.T. 2009).

Most present day inhabitants of the South-Central region are descended from migrants, so they pay great reverence to their ancestors, with village founders worshipped as *thanh hoang* (Tutelary Deities). In general, each village has one communal hall, which consists of a complex of five components, each with different religious functions (Table 6.1).

The moral basis of Vietnamese society is anchored in the tradition of 'remembering the source from which one drinks water', an expression of the deep sense of gratitude to the ancestors for their labors and struggle to survive and build a prospering community (Truong n.d.). In earlier times the appeasement of a mysterious and often hostile natural environment was also important. This led to a strong belief in the power and salvation of numerous deities. Therefore, when the economic basis of a new settlement had been established, villagers constructed a shrine for the village's tutelary genie, ancestral sages and wise elders. The shrine became the principal cultural and

Photo 6.3 The van chai building and market at Lach Vinh Troung, Nhatrang, Khanh Hoa Province, Vietnam

Table 6.1 The five components of fishing village communal halls in the South-Central Region

	Vernacular	English	Function
1	*dinh trung – chanh dien*	Hall center – main palace	Worship the Tutelary Deities
2	*tien hien*	Early men of merit house	Worship the first contributors to the construction of the temple
3	*tien sư*	Deceased predecessors	Worship the ancestors of the fishers
4	*lang ba*	Goddesses' mausoleum	Worship goddesses of the five elements; *kim* (metal), *moc* (wood), *thuy* (water), *hoa* (fire), and *tho* (earth)
5	*dai khanh*	Great happiness house	Meetings of village elders

Photo 6.4 Leng Ong Thuy (Altar for the Male Water God), Van Thuy Tu, Phan Thiet City, Binh Thuan Province, Vietnam

organizational center of a village (Nguyen T.C. 1993; Claeys 1943; Huard and Durand 1954; Lam 1996; Huynh 1996; Nguyen X.L. 1996; Ruddle 1998) (Photo 6.4).

From the late-eighteenth century marine fishing villages enlarged their shrines for the worship of the Deity of the South Sea. This deity is a 'whale' (a local concept that embraces all cetaceans). The 'Whale Shrine' became the locus of moral authority of a fishing community's life, and the foundation on which fisheries management was and remains based (Ruddle 1998). According to Claeys (1943), the cult of 'whale' veneration originated during the reign of Emperor Gia Long. In gratitude for reputedly having been saved from disaster by a whale while at sea near Phu Quoc Island, in 1783 Emperor Gia Long elevated whales to a high grade mandarinate. Supernatural powers were attributed to cetaceans, and they became venerated.

South of the South-Central coast the modern functions and formalities of van chai are substantially different from those along the Northern coast of Vietnam.

Whale temples are traditional institutions where fishers worship their marine gods, together with their ancestors who developed and managed the community's fisheries. Most temples in the South-Central provinces are dedicated to the whale, and ceremonies are conducted there to venerate it, since fishers believe that the whale is the deity which protects men at sea.[9]

6.4.1 The Structure of 'Guild Type' Van Chai-based Fisheries Management

Although the details vary considerably by locality, the underlying principles of the veneration of deities and ancestors, combined with the sacred obligations of mutual assistance, remain all pervasive. The sections that follow are based on the comprehensive data from Van Thuy Tu.[10]

6.4.1.1 Van Administration

An elderly man of high prestige and with a profound understanding of local society and fishing usually heads a van chai. At Van Thuy Tu the administrative committee is composed of 7–15 members, elected to a three-year term of office. All boat owners and fishers 18 years-of-age or older elect them, and all over 21 years old can run for election. The elected administrators themselves elect the three heads of the sub-sections of administration that manage routine affairs. These are the Head of Worship, Head of the Van, and Secretary of the Van.

Van administrators have four main duties. These are: (1) worship of the Sea Deities; (2) assisting the local government to implement the orders of higher levels of government; (3) in concert with the Hamlet Council, settling fisheries disputes among fishers; and (4) investigating the needs of the fishing community and assisting the government to fulfill them.

6.4.1.2 Mutual Assistance

A major function of van administrators is the maintenance of the shrine and conduct of festivals and routine ritual performance. The number and scheduling of

---

[9]These have been described in details elsewhere (see Nguyen, D.T. 2009; Ruddle and Luong 2009; Tran 2009).

[10]Because of the far-reaching changes occurring in both general society and the fishery, in 1963 officials of Duc Thang Ward decided to ensure that local fishery regulations and customs be handed-down systematically to future generations (Republic of Vietnam 1963). To do this they compiled a comprehensive document consisting of 22 chapters and 114 articles dealing exclusively with the local marine fishery (Chau Thanh Village Council 1963). That document is analyzed here.

shrine festivals varies by van, but most are held in the Spring and Fall, according to lunar reckoning (Nguyen 2002a; Nguyen D.T. 2009).

The linkage between festivals and mutual assistance in the codification document of Van Thuy Tu implies that mutual assistance is a sacred duty of van members (Republic of Vietnam 1963). It demonstrates the traditional moral authority of the van. It is reiterated within the document that the Sea Deities must be solemnly and sincerely worshipped by owners of fishing boats and fishers.

The rituals performed at these festivals are of deep significance, since they emphasize the importance of harmonious relationships among the various stakeholders in the fishery, and the vital importance of mutual assistance, both of which ensure the continued prosperity of the community and thereby venerate the ancestors. For example, the centerpiece of the 'Praying for Fish Festival', held in the seventh lunar month, is the *Ba trao* traditional operetta. This morality play interweaves three themes stressing the community's traditional values. These are (1) that by combining the intercession of supernatural forces with their own diligence and sacrifice, the community's ancestors could create a vibrant economic and social life under harsh frontier zone conditions; (2) the life of a fisher is hard and dangerous, so if the fishing community's aspiration for a comfortable and prosperous life is to be realized, a spirit of unity, attachment, benevolence, righteousness and affection among all persons in the fishery is essential; and (3) the benevolence of the 'whale' saves human life at sea (Ruddle 1998).

Together, these three themes sanction the moral authority of the van. In particular they emphasize especially the moral obligation for mutual assistance and community solidarity (Republic of Vietnam 1963).

Mutual assistance obligations are specified in detail. They are divided into three groups of rules: (1) those applied at sea, (2) those pertaining to a bereaved fishing family, and (3) those pertaining to substitute crewmembers. If an accident occurs at sea, boat owners and fishing boat crews must try first to save the crew, and then the boat and gear. Those who do not would be punished by the court. When requested by a boat owner the van organizes emergency searches for accident victims. If a fisher dies at sea during the fishing season, the van must try to supplement government assistance to the bereaved family. The family is allocated food, boat owners are enjoined from claiming the dead fishers' loans or advances, and the victims' parents, wife or children are entitled to his share of catch proceeds. The boat owner pays the funeral expenses. To maintain a proper crew size, boat owners are obliged to seek substitute members for those lost at sea. The boat owner is required to pay their hiring expenses (i.e., wages, advances or loans) (Ruddle 1998).

6.5 The Design Principles of Van Chai

The 'design principles' that characterize the van chai comprise use rights, various sets of detailed rules that govern how those rights are exercised legitimately and by whom, monitoring and accountability regarding the rules, conflict resolution that governs settlement of disputes if rules are broken, and punishment that is applied to those who break the rules. In the case of the van chai the predominant design principles are those reflecting and governing human relationships among the various stakeholders in the fishing community, in the context of mutual assistance and respect as governed by the precepts of the whale shrine.

6.5.1 Rights

Under pre-existing systems such as the van chai resource use is governed by use rights protected by customary law and practice. Such a grant of authority defines legitimate uses as well as the penalties for violating those rights. In Vietnam the main rights are primary or birthrights, residential proximity rights, the right to sell, lease or bequeath the right, and that to share rights.

Fishing rights in Tam Giang lagoon of Thua Thien Hue Province historically took two forms (Anon 1994). In the first, the Central government authorized the local (district, commune) governments to auction fisheries and collect taxes. In the second, a Royal Bonus was awarded to villages of outstanding merit, or in some cases to a military unit in wartime, or to a specific geographical area (water surface). The villages would then have to pay tax, but they were allowed to fish by themselves or with other villages or individuals with their permits. They could transfer such rights to other villages, but sale to private individuals was prohibited.

The host villages (or tender winners licensed by the State to collect tax) would accordingly convene meetings for all van chai in the region, to evaluate them and determine their tax quotas. Each van chai then held a follow-up meeting(s) to evaluate specific occupations and locations in order to impose tax levels on fishing households or individuals. Besides the sharing of fishing rights and tax responsibilities, the van chai may have had other regulations and rules about coordination and management of fishing grounds and common resources, designed to avoid disputes and conflicts (Nguyen 2002b).

6.5.1.1 Primary Rights

Most commonly these are a birthright. In the Dong Hoi area of Quang Binh Province, for example, a person who inherits his or her father's residence also inherits the associated proximate fishing right (Ruddle 1998). Sons of an in-marrying male could inherit bilaterally, if their father also inherited rights elsewhere, from his father. Again this varies throughout the country. At Van Son Hai, in Quang Nam Danang Province, for example, only males born in the village and who continue to reside there inherit the right to fish in the village's exclusive sea territory. Outsiders must wait for a minimum of ten years before being granted fishing rights, so they fully understand the village's 140-year old traditional fishing rules and behavior. Further, a man born into a fishing family but who had moved away for a protracted period could not automatically expect to enjoy his fishing rights should he return to the village later in life. Unwritten rules state that acquisition of rights depends on a person having built-up adequate local experience before receiving a personal right (Ruddle 1998).

6.5.1.2 Right Conveyed by Proximity to Residence

A fundamental but not nationwide right governing coastal fisheries management is that to operate small-scale fixed gear in waters proximate to one's residence. In the Dong Hoi area of Quang Binh Province, for example, fishing spots for fixed gear (such as those suitable for employing a lift net to catch small pelagics) near a one's residence can be claimed exclusively by the householder. But in other areas this right can be over-ridden by the 'first-comer rule' (see below). For example, at Van Son Hai, in Quang Nam Danang Province, the first person to erect fixed gear, even if adjacent to another person's residence, enjoys first comer's user rights until he dismantles the gear (Ruddle 1998).

6.5.1.3 The Right of Transfer and Loan

In pre-colonial times, local governments were empowered by the kings to lease via auction and tax local fisheries. Since under these arrangements families with a history of fishing were given priority in obtaining leases, many fishing grounds became a quasi-private property of a fishing family, with the subsequent inter-generational transfer of rights resembling inheritance. Meritorious individual villagers were also granted property rights to the fishery. However, these rights were taxed, and did not include the right of sale. Centralized governance was weak under indigenous Vietnamese rule, so local communities essentially managed the fishery (Ruddle 1998).

In the Dong Hoi area of Quang Binh Province rights of residential proximity can be loaned, sold, leased, given, or otherwise transferred, either permanently or for a fixed term. Where a person transfers his right either permanently or for the long-term to another, who then sells the residence to a third person, the person buying the house can also claim the right of residential proximity. However, the original owner could negotiate the return of the right from the third purchaser of the house. At Van Son Hai, Quang Nam Danang Province, fishers are permitted to loan their annually re-allocated rights to another fisher from the same village (Ruddle 1998).

6.5.1.4 Shared Rights

There is little information on the sharing of fishing rights in Vietnam, but the practice was apparently traditional and widespread. In the Dong Hoi area of Quang Binh Province, for example, residential proximity rights are shared when they intersect or overlap. Families there also make agreements to share residential rights with those lacking them. At Van Son Hai, Quang Nam Danang Province, fishers are permitted to share their annually re-allocated rights with others from the same village (Ruddle 1998).

6.5.2 Rules

6.5.2.1 First-Comer's Rule

Common throughout Vietnam is that on the open sea the right of a first-comer to the exclusive use of a fishing spot is always upheld, regardless of the gear type being employed (Ruddle 1998).

In a 'floating village' van chai there is a priority rule for first-comers. Normally, the priority rule has the following contents (Nguyen 2002b):

(a) If the fishing activities take place according to the direction of fish movement, later-comers have to drop their nets behind those of the first-comers, and they may neither impede nor interfere in any way with the first-comers' nets.
(b) When fishers discover fish within a fishing ground, the first-comers may drop their net to catch the fish in their direction of movement, but latecomers must place their nets in a correct relationship to those of the first-comers.
(c) The allocation of places for boat mooring follows specific rules that establish their order of mooring. According to this order, the van's boats are allowed to capture fish at certain times during the day. In other words, the boats take turns to cast and haul their nets according their arranged mooring order and at the times defined.

Fishers using nets in the van chai on the Bang River (Cao Bang Province) change their fishing order frequently. A man who was first at an earlier time goes to the end of the row of boats to fish last the following time, and so on.

Under those circumstances, and where there are too many fishers, many later-comers would not be able to catch fish. Although access is free, and outsiders may ask for access, if a van is already crowded new members would never receive an opportunity to fish, as they always have to wait at the end of the boat row in the van's locations. If an existing van develops rapidly and the membership becomes high (owing to the new young families living apart from their parents), some members leave to find new fishing grounds, where they establish a new van. As a result the number of residents of each van in rivers or lagoons remained relatively stable

6.5.2.2 The Definition of Fishing Territories

Exclusive village sea territories were widespread prior to and during French colonial times. A village's sea territory was usually, but not always, defined by proximity or adjacency to its settlement, and by lateral and seawards boundaries. The depth or other limits at which gear could be operated defined seaward boundaries. The village elders fixed these boundaries (Ruddle 1998).

6.5.2.3 Inter-community Access Rules

Access controls were applied to outsiders. Commonly, rules specified that some form of fee, compensation or royalty be paid once permission has been granted. At Van Soi Hai, Quang Nam Danang Province, for example, fishers from other communities have traditionally been permitted to fish within the village's exclusive sea territory in return for payment equivalent to 30% of the value of the catch caught by them in the village's waters (Ruddle 1998).

6.5.2.4 Gear Rules

Gear rules are widespread (Ruddle 1998). Many were established to overcome gear externality problems. In Tam Giang Lagoon, Thua Thien Hue Province, the emphasis on overcoming gear externalities and ensuring equity of catching opportunity is well illustrated (Nguyen X.L. 1996). There minimum distances were specified between fish corrals (1 km), and between fish corrals and bottom gill nets (1.2–1.5 km). Further, to ensure equity bottom gill nets could not be set at the lagoon entrance and fixed gear must be set at least 10 m from the edge of the lagoon.

However, as at Van Thuy Tu, where detailed rules were applied to the 11 main gear types used, rules pertained mostly to eligibility, seasonality and profit-sharing among boat owners, captains and crew. The overriding principle of eligibility rules is that boat owners are eligible to participate in a given fishery if they can supply the requisite number of boats and gear, as well as meet all expenses for the entire season. The details of profit sharing among boat owners, captains and crewmembers vary by gear type. They are summarized in Table 6.2. Further detailed rules are applied to some gears, and particularly to the fixed Sardine net (Table 6.3).

6.5.2.5 Temporal Allocation Rules

In general, rules are enforced to promote both orderly and equitable fishing. These can be both short- and long-term. At Van Son Hai, Quang Nam Danang Province, for example, before the opening of each fishing season all village fishers gather for an annual meeting, to allocate exclusive fishing spots within the village sea territory (Ruddle 1998). At that time the head of the van allocates to groups of fishing units their exclusive rights areas for the next year. These groups must pay 20–30% of the profit from their catch to the van. The grounds are allocated to all fishers, regardless of gear type operated, and verified by a vote by all fishers at the meeting. During the second lunar month, just prior to the annual meeting, the approximately 50 named

Table 6.2 Profit sharing Rules by gear type at Van Thuy Tu, Binh Thuan Province

Gear type	Profit sharing
Gizzard shad net	50–50 owner and captain-crew; captain of forward boat 2 shares, captain of rear boat 1.5 shares, crew 1 share each
Drag net	Two types: 50–50 owner and captain-crew, and 40–60, respectively. Captain of forward boat 2 parts and of rear boat 1.5 parts
Small drag net	30–70 owner-captain and crew. Captain takes 1.5 shared, crew 1 share
Trawl	50–50 owner-captain and crew. Captain gets 1.5 shares, crew divide balance equally
Purse seine	40–60, owner-captain–crew
Sea crab net	Each member keeps income from own net; boat owner gives 10% of his receipts to helper (who has no gear); crew give 10% of receipts to boat owner and 10% to helper
Nylon net	Each person keeps income from own net and pays 15% of receipts to boat owner
Shell-fishing	Equal division among boat owner and two cooperating crewmembers
Hand-lining	Boat owner takes 15% of receipts from each day's catch
Fixed gear	Each participant entitled to receipts from own gear

Ruddle (1998), based on field survey data obtained in 1995–1996

village fishing grounds are classified into three quality categories, based on water depth. These are 'Best', defining inshore waters to a depth of 15–18 m; 'Intermediate', referring to waters of 18–25 m depth; and 'Worst', meaning 'pockets' of deeper water that occur along the shoreline. To ensure equitable treatment, fishing units are annually reallocated grounds of different quality on a rotational basis. The allocation is exclusive for the year. The approximately 50 grounds are allocated among the almost 400 units in the village, so that an average of eight units is allowed to operate per ground. But the precise number varies according to the quality of the ground.

The head of the van has no authority of closure. Instead, the fishers holding annual exclusive rights to a particular ground do this. The head's authority extends only to areal allocation of sea space.

6.5.2.6 Conservation Rules

Compared with other aspects of fisheries management, resource conservation appears to have been of little concern in the traditional van system, since relatively few traditional rules appear to have been directed it. According to informants, resource depletion is a recent concern resulting from motorization and the use of more efficient gear by a greatly increased number of coastal fishing units (Ruddle 1998).

This same situation has also been observed in Tam Giang Lagoon, Thua Thien Hue Province. Although in former times the fishing communities had no strict rules for resource protection, a minimum mesh size of 1 cm for cotton gill nets and a minimum spacing of 1.5 cm between the bamboo poles of fish corrals permitted

Table 6.3 Rules applied specifically to the sardine net at Van Thuy Tu (Chau Thanh Village Council 1963; Ruddle 1998)

Object	Rule
Eligibility	Only boats that constructed a fish shelter and set the net could fish, unless the owner had permitted use of the site (see below). To be eligible a boat must construct at least one shelter.
Fishing season	The season opened from the first lunar month, except for late-entrants who begin in the fifth month, and ended on the fifth day of the ninth month, when the season's income was shared. If abundant fish remained and sea and weather conditions were still favorable, fishing could continue. The same sharing rules as for the main season governed the income from this extended season.
Territory	To avoid gear conflict, individual nets must separated by at least 300 m. Net owners had to mark their sites with a distinctive sign. Users of other gear, particularly nylon nets, seine nets and baited-lines, were forbidden from fishing around the sardine nets belonging to others.
Conservation	Attractor lights were forbidden because, although a large catch would be made on the first night, thereafter few fish could be caught. Blast fishing was prohibited.
Outsider's rights	Boats that had set their own Sardine net had the right to fish from another's site. However, as soon as the boat that 'owned' the site approached and signaled, an outsider boat had to cease fishing immediately. Should the outsider fail to comply immediately, and delay the owner's fishing, the latter could claim the entire catch from the outsider. However, the owner was forbidden to take any direct punitive action at sea.
Removal of shelters	So the entire community could benefit, rules governed the removal of shelters at the end of the fishing season. Fishers could remove only their own fish shelter; shelters could not be removed before the 21st day of the 9th lunar month, in case others wished to continue fishing after the season closure, on the 5th day; and fishers who stopped fishing before the end of the season had to leave their shelters intact until the 21st day of the 9th month, for the use of others.

small fish to escape. But these rules no longer apply now that nylon gill nets are used and corrals are made of netting. Blast fishing and using stupefacient substances derived from plants were also prohibited (Nguyen X.L. 1996; Tran 1996b). Nowadays, fishers often strictly monitor each other to ensure that nobody uses either electrical gear or dynamite for fishing. Violators are sent to the police for punishment (Ruddle 1998).

6.5.2.7 Distribution of Catch Rules

In Vietnam rules defining remuneration of harvesting labor are complex and vary greatly by fishing port and gear type. But everywhere it is based on a share of the annual catch value, after all costs have been deducted, divided among boat owners, captains and crewmembers. The information provided here is derived entirely from Van Thuy Tu (Table 6.3). Boat owners, captains and crew-members are entitled to sell the catch at the landing site for prices that are published daily.

6.5.2.8 Rules Pertaining to Relationships Among Boat Owners, Captains and Crew

A fundamentally important set of rules defined relationships among boat owners, captains and crewmembers. These are strenuously enforced to ensure harmony in the fishing community. Operational rules govern fishing behavior, gear externalities, assignment issues, fishing behavior and temporal allocation of space, seasonality of fishing, conservation practices, and distribution of the catch within the community (Ruddle 1998).

Several rules cover boat owners' behavior when seeking to hire a captain and crew. First they are required to know for whom the captain and crewmembers worked during the previous season. They are forbidden to offer enticements to lure men away from other boat owners, and must not compete to hire a captain and crew.

Formerly at Van Thuy Tu all contractual rules regarding the rights and obligations of crewmembers concerned cash advances and loans from the boat owner. Such financial arrangements are still made, but maximum amounts allowable are no longer set by van rules (Ruddle 1998). Contracts specify the following three types of financial arrangement. (1) Before starting work for the year, crewmembers are entitled to an interest-free cash advance to cover their family's expenses. This is repaid by deduction from each crewmember's share of the profit from the entire fishing season. In a poor season, where the individual crewmember's share of the profit is less than the amount advanced, repayment must be made by either signing-on with the same boat owner for the following year, or by obtaining a loan from another boat owner to repay the advance in full. (2) Crewmembers can obtain an interest-free cash loan from a boat owner who is unable to make an advance. Such loans must be repaid in cash, but are not subject to deduction from shares of the fishing season's profit. (3) Crewmembers who receive advances or loans, but who fail to fulfill their contracted duties, are required to make repayment in full. They are not entitled to wages. A captain's rights and obligations to loans and advances are the same as for crewmembers, except that the maximum amounts allowed are double.

Captains and crewmembers are admonished to cooperate closely in the conduct of fishing operations. Those who become drunk then argue and fight, or who quit fishing during the contract period, and thus interrupt smooth fishing operations, are taken to the authorities for punishment. The same rule is applied to crewmembers who importune several boat owners for loans, as well as to 'stubborn and bad-mannered captains and crewmembers' (Chau Thanh Village Council 1963).

Captains and crewmembers are also required to maintain gear. If intentional damage to gear can be proven the expense of repairs is covered temporarily by the van, and later deducted from the profit share of the guilty person(s). Such persons may also be subject to criminal proceedings.

Captains and crewmembers are required to work for the entire fishing season. However, despite having signed an annual contract, early resignation can be accepted. In such a case they must seek an acceptable substitute person, and also repay all advances and loans. If not they could be tried, and forced to compensate the boat owner, captain and fellow crewmembers.

Provision is made for illness. Captains and crewmembers are entitled to five days sick leave per season, with an additional five days if the illness is both serious and clearly work-related. Under normal sea conditions, other crewmembers will cover for a sick colleague for up to ten days. However, when seas are heavy, the van is required to hire a substitute. If a person is on sick leave for more than ten days, he must himself hire a substitute, in order to remain entitled to his full share of the season's profit.

6.5.2.9 Operational Rules

Detailed rules were applied to the main gear types used. These pertained mostly to eligibility, seasonality and profit sharing among boat owners, captains and crew. The overriding principle of eligibility rules is that boat owners are eligible to participate in a given fishery if they can supply the requisite number of boats and gear, as well as meet all expenses for the entire season. The details of profit sharing among boat owners, captains and crewmembers vary by gear type (Table 6.2).

Further detailed rules are applied to some gears, and particularly to the fixed sardine net. At Van Thuy Tu special rules were applied to the sardine net, because, being a major fishery and employing a fixed fish sheltering device, there was a greater inherent potential for conflict than with other gear types. The sardine net comprised a fixed floating fish shelter made of bamboo, branches and palm fronds, and anchored with ropes and rocks. Since this gear targeted an important fishery (sardine was the basis of the local fermented fish sauce [*nuoc mam*] industry), it was governed by an elaborate and specific set of rules. These pertained to eligibility, fishing season, territory, conservation, outsider's rights, removal of fish shelters, monitoring, expenses, accounting, and catch-sharing.

6.5.3 Monitoring and Accountability

If rights are to be meaningful, provision must be made within the system for monitoring compliance with rules, and back-up this by imposing sanctions on violators. At Van Thuy Tu the fishers monitored the sardine net fishery and were required to report to the van officers any violations concerning the location of fish shelters or lack of maintenance.

A detailed contract had to be drawn-up at the beginning of the fishing season at a special meeting of captain and crewmembers. Details of the season's work assignment, rewards, punishments, rations, and the like had to be specified, and the agreement recorded in a minutes of the meeting. A detailed record of expenses and other accounts was required. To avoid conflict, each fishing unit had to employ a bookkeeper, who could be neither a family member nor relative of the boat owner, captain or any crewmember, to maintain clear and impartial accounts for the season. The bookkeeper was paid with 3% of the season's profits, as was the person in charge of selling fish and collecting sales money. As a further safeguard, the boat owner and crewmembers had kept an account book for comparison.

In the community of Cau Hai, at Tam Giang Lagoon, Thua Thien Hue Province, the head of the van was assisted by a designated group of 'guards'. Nightly they patrolled the waters for fishers using either dynamite or stupefacient substances. Any such violator was apprehended and judged by the head of the van (Nguyen X.L. 1996).

6.5.4 Conflict Resolution

Gear or other kinds of conflict among fishing units is not uncommon, particularly where trawling and fixed gears are the major techniques. In Vietnam there are three stages to conflict resolution: First is settlement among the fishers themselves (most conflicts are settled in the way); then resolution by the van Committee; and finally resolution by the People's Committee.

Gear conflict or infringement of first-comer's rights and rights of residential proximity is still generally resolved in the fishing community by the village elders. But nowadays if they cannot resolve their problems they are taken before the People's Committee. (In former times problems would have been taken to the Village Magistrate.)

Violations of Van Thuy Tu fisheries regulations are first considered by the Administrative Section. It gives its opinion to the Hamlet Council, which then settles matters and imposes punishment according to statutes regarding local government. The objective is to mete-out impartial and constructive punishment that provides an example to the community.

The principal officially sanctioned authority vested in the Administrative Section of a van is its power to conciliate fisheries conflicts, the resolution of which is not stipulated in current local rules or higher laws. However, its power is limited in that it can be exercised only at the request of either boat owners or fishers. The Administrative Section forwards impartial opinions to the Hamlet Council to assist the latter in making decisions. The powers of the van administrative section are tightly circumscribed; it cannot make a decision, being limited to an advisory role only, and its conciliation efforts must be requested by plaintiffs. It alone cannot initiate, implement or uphold new rules or local laws.

In conformity with the concept of constructive sanctions, punishment is graduated in severity. Sanctions are applied at four levels: first is explanation and seeking the wrongdoer's understanding. Then a warning is given. The third level is a monetary fine by the Hamlet Administrative Section. Finally the most severe punishment is revocation of the fishing license by the provincial authorities acting on the advice of the Hamlet Council.

Interpersonal disputes within an individual fishing unit are handled differently. As is clearly set forth in the codification document of Van Thuy Tu, mutual respect for the rights and dignity of all persons involved in a fishing unit must be upheld. Boat owners are forbidden to beat or humiliate captains and crewmembers, and vice versa. Violators are punished according to the national penal code.

6.5.5 Sanctions

As throughout the Asia-Pacific Region sanctions are widely invoked in Vietnam for the infringement of fisheries rights and the breaking or ignoring of locally formulated rules governing fishing and marine resource use. These days either social or economic sanctions are applied. Implicit for those fishers whose belief in the moral authority underlying traditional shrine-based management remains strong is that failure to abide by locally made rules, particularly those pertaining to mutual assistance, would invite supernatural sanction in the form of hazards at sea (Ruddle 1998).

However Tran mentions that at Vinh Ha, Tam Giang Lagoon, Hue Province, those who fished in others' rights areas were punished either by public criticism or by being ostracized by their community (Tran 1996b). In contrast, these days economic sanctions are more commonly meted-out. Crew-members in cases involving deliberate gear damage or destruction are punished by being made to pay the claimant compensation equivalent to the amount of the loss. Tran notes that at Vinh Ha, fishers using mobile gear who were either late in paying or failed to pay their taxes had either their gear or catch confiscated (Tran 1996b).

6.6 Concluding Comments

Pre-existing systems like the van chai often change rapidly, in response to a wide range of both external stresses and internal pressures. In Vietnam, a complex and turbulent political history has had a major impact on pre-existing management systems. Another major cause of change was motorization of fishing vessels and gear introductions, such that use of fixed gear declined concomitant with the introduction of mobile techniques, especially purse seining and trawling. As a result, in many locations time-honored rules that formerly governed fisheries have lapsed.

Despite years of turmoil, in many areas the systems have survived. The core of the van chai system has proven remarkably resilient, undoubtedly because the salient characteristic of pre-existing management systems in Vietnam is regulation of interrelationships among fisheries stakeholders, within the framework of the strong moral authority of the community shrine, rather than regulation of fishing and the fishery per se.

Thus in the many coastal fishing villages of Binh Thuan Province van chai have been maintained by the community, since they are important both in fisheries production and in social and spiritual life. Similarly, van chai have managed fishery activities in Tam Giang Lagoon, Thua Thien Hue Province since feudal times. The selective use of traditional village values is regarded as one potential way of managing the present day fishery there (Nguyen Q.V.B. 2009).

Following the colonial era, the successor governments of the independent Vietnams paid no attention to the pre-existing management systems, and in the provinces that comprised the former Democratic Republic of Vietnam, as well as at various locations in the Central and Southern regions, the traditional religious characteristics have lapsed, and only the secular administrative functions remain. During the French colonial era religious functions were still performed in the Northern provinces. For example, Claeys (1943) noted that at the fishing community of Li Hoa, just north of Dong Hoi, Quang Binh Province, a major festival was observed at the whale shrine during the fourth lunar month. But nowadays each fishing community in Quang Binh Province has only a 'Problem-Solving Committee', which, in addition to solving fisheries-related problems, also addresses social issues within the village. Nothing else remains of former systems.

In contrast, at Van Lach Thang Tan, near Vung Tau City, Ba Ria Vung Tau Province, for example, the van chai now retains only its religious functions, and has been developed as a tourist attraction (Photo 6.5). The Ong Nam Hai shrine in this van was officially established in 1802, and greatly elaborated in 1824, when the population increased sharply after an influx of settlers from the central provinces. Documents permitting the establishment of the shrine and confirming the fisheries rights of the van chai were issued in the Minh (1820–1840) and Thieu-tri (1841–1847) eras. However, fisheries regulations lapsed in the late-1940s, when the provincial government established an open access system for provincial vessels. Since they were never documented, knowledge about them was gradually forgotten as the system fell into disuse (Ruddle 1998).

Photo 6.5 A whale shrine renovated for tourism at Van Lach Thang Tam, Ba Ria Vung Tau Province, Vietnam

Then, following national reunification, fishing vessels were put under public ownership, during the Collectivization Period (1975–1988). Fishery cooperatives were developed by the State (Le and Tuong 2009), so the role and operations of van chai was overshadowed, and many van temples were neglected.

Investment in the sector was renewed during the Market-Oriented Economy Period (1988 to present), with the implementation of policies to revive fisheries. However, after a long period of idleness during the preceding Collectivization Period many van chai had become derelict. In recent years the Government has made many attempts to preserve or revive national cultural identity, based on van chai. This has included support for rebuilding or repairing Whale temples, and re-organization of fishing community festivals Tran (Ruddle and Tuong 2009; Tran 2009). As a result, many large and key temples have been reconstructed, and gradually the role of van chai has revived. The fishing communities welcomed such positive changes. However, so far the activities of van chai have focused mainly on cultural and religious traditions, rather than on their important roles in community cohesion and in the organization, management and development of fisheries production and resource protection.

It goes without saying that every social entity emerged in accordance with its own innate production capacity in terms of prevailing cultural and social standards. The centuries old van chai and fishing community system of Vietnam was established in locations with a small and local demand for fish, was shaped by small population groups under conditions of little pressure on aquatic resources, and based on small-scale and simple but sophisticated means of production. Such fishing communities usually existed independently, and were little affected by forces outside their immediate, local social system. In social terms, the van chai was based on the principle of mutual assistance or 'neighborly affection' (*tinh lang nghia xom*), and rooted in ethical and behavioral standards based on the Confucianism all pervasive in Vietnamese culture. Satisfaction of the spiritual needs of the fisher and his community was of fundamental importance, and was among the main functions of a van chai.

However, all this has changed in fundamental ways over the years. The implication is that if the van chai is to play a role in the administration of modern fisheries and fishing communities, its underlying principles must be adapted and applied within an entirely different framework than the one under which it arose initially. In particular, it must become both integrated within a larger administrative framework and locally embrace more than just fisheries.

References

Anon. (1988). *Dictionary of the Vietnamese language*. Hanoi: Social Publishing House.

Anon. (1994). *Lagoons of Thua Thien Hue: Proceedings of the workshop on research project coded 03-11 under State Program KT-03*. Hai Phong: Haiphong Sub-Institute of Oceanography.

Anon. (1996). *Van Thuy Tu temple profile* (unpublished report). Documentation on national cultural relicts proposal. Phan Thiet: Binh Thuan People's Committee (in Vietnamese).

Chau Thanh Village Council. (1963). *On the compilation of village regulations*, Official Note No. 982/ACL (May 17, 1963). Phan Thiet: Chau Thanh Village Council (in Vietnamese).

Claeys, J. Y. (1943). L'Annamite devant la mer. *Indochine, 3*, 1–10.
Dao, D. (1938). *Outline of the cultural history of Vietnam*. Hanoi: Bon Phuong Publishing House (in Vietnamese).
Ha, X. T. (1995). *Overview of fishing villages in Vietnam*. Hanoi: Institute of Fisheries Economics and Planning.
Ha, X. T. (1998). Traditional organisation of fisheries communities in Vietnam, volume I. In *Proceedings, IIFET'98 Tromso* (pp. 286–292). Corvallis, OR: International Institute for Fisheries Economics and Trade.
Ha, X. T., & Nguyen, D. T. (2009). The *van chai* and its role in the hierarchy of fisheries administration in Vietnam. In K. Ruddle & P. L. Tuong (Eds.), *The van chai of Vietnam: Managing nearshore fisheries and fishing communities* (pp. 21–34). Hong Kong: International Resources Management Institute.
Hoang, T. V. M. (2003). *The cultural factors of dwellers of My Van (Hong Ha commune, Dan Phuong District, Ha Tay Province)*. Hanoi: History Faculty. Social and Human University (in Vietnamese).
Huard, P., & Durand, M. (1954). *Connaissance du Viet-Nam*. Hanoi: Ecole Française d'Extreme-Orient.
Huynh H. L. (1996). Van Thuy Tu in Phan Thiet. *Big Cooperation, 63*, 4 (in Vietnamese).
Lam T. T. (1996). Van Thuy Tu, Artistic architecture and popular religion by the seashore of Phan Thiet. *Culture and Art: Phan Thiet, 18*, 19 (in Vietnamese).
Le, T. L., & Tuong, P. L. (2009). A comparative analysis of fisheries management via pre-existing *van chai* and contemporary cooperatives. In K. Ruddle & P. L. Tuong (Eds.), *The van chai of Vietnam: Managing nearshore fisheries and fishing communities* (pp. 86–97). Hong Kong: International Resources Management Institute.
Nguyen, D. T. (1984). *Bo Lo* people and *v n Ky Xuyen*. *Ethnology Magazine, 3*, 44–49 (in Vietnamese).
Nguyen, D. T. (1993a). A few features on the way of life of the Bo Lo and other fishing inhabitants at Cua Sot estuary. *Vietnam Social Sciences, 4*, 158–211.
Nguyen, D. T. (2002a). Religions and beliefs in Vietnamese fishing communities. *Ethnology Magazine, 1*, 62–71.
Nguyen, D. T. (2002b). *Fishing communities in Vietnam*. Hanoi: Social Science Publishing House (in Vietnamese).
Nguyen, D. T. (2003). Sea daily routes of the Bo Lo in the Northern area of the central region. *Magazine Folklore Culture Review, 6*, 27–32 (in Vietnamese).
Nguyen, D. T. (2007). Studies of fishing communities in Vietnam. *Anthropology Review, 2*, 45–57.
Nguyen, D. T. (2009a). The role of floating fishing villages in the social life of fishers in the South-Central Region. In K. Ruddle & P. L. Tuong (Eds.), *The van chai of Vietnam: Managing nearshore fisheries and fishing communities* (pp. 36–46). Hong Kong: International Resources Management Institute.
Nguyen, Q. T. T. (1995). *Vietnamese fisheries in the first half of the 20th century*. Hue: Thuan Hoa Publishers.
Nguyen, Q. V. B. (2009b). *Van chai* in Thua Thien Hue: Tradition, present and future challenges. In K. Ruddle & P. L. Tuong (Eds.), *The van chai of Vietnam: Managing nearshore fisheries and fishing communities* (pp. 66–76). Hong Kong: International Resources Management Institute.
Nguyen T. C. (1993). The traditional Viet Village in Bac Bo: Its organizational structure and problems. In: H. L. Phan, T.C. Nguyen, D. N. Nguyen, K. Q. Duong, V. B. Cao, D. D. Phan, V. Huy, L. To, K. T. Nguyen, D. N. Nguyen,T. Chuong, X. N. Pham, & S. Nguyen (Eds.), *The traditional village in Vietnam* (pp. 44–142). Hanoi: The GIOI.
Nguyen, X. L. (1996b). Van Thuy Tu: Popular religion near the seashore. *Phan Thiet Spring Magazine*, 22–23 (in Vietnamese).
Republic of Vietnam. (1963). *Minutes: Seminar on amendment to professional rules concerning the sea fishery of Van Thuy Tu. Phan Thiet: retained by Duc Thang Ward*. Phan Thiet: Duc Thang Ward Office (in Vietnamese).
Ruddle, K. (1998). Traditional community-based coastal marine fisheries management in Viet Nam. *Ocean and Coastal Management, 40*(3), 1–22.

Ruddle, K., & Luong, T. S. (2009). *Van chai* in fisheries village development and management in Binh Thuan Province. In K. Ruddle & P. L. Tuong (Eds.), *The van chai of Vietnam: Managing near-shore fisheries and fishing communities* (pp. 47–65). Hong Kong: International Resources Management Institute.

Ruddle, K., & Tuong P. L. (2009). The importance of pre-existing local management systems and their context in Vietnam. In K. Ruddle, & P. L. Tuong (Eds.), *The van chai of Vietnam: Managing nearshore fisheries and fishing communities* (pp. 1–20). Hong Kong: International Resources Management Institute.

(SARI) Southeast Asia Research Institute. (1998). *Sea and the ancient Vietnamese*. Hanoi: Culture Publishing House (in Vietnamese).

Tran V. T. (1996a). *Unwritten regulations of Fishers in Vinh Ha Commune*. Report of Biological Resources of Tam Giang Lagoon Research Project. Hue: Hue Agriculture University.

Tran, V. T. (1996b). Informal rules in the fishing community of Vinh Ha. In B. Brzeski & V. T. Truong (Eds.), *Annual technical report (first year)* (pp. 60–62). Hue: Centre for the Management of Biological Resources in Tam Giang Lagoon.

Tran, V. V. (2009). Whales in the spiritual life of coastal fishing villages in Binh Dinh Province. In K. Ruddle & P. L. Tuong (Eds.), *The van chai of Vietnam: Managing near-shore fisheries and fishing communities* (pp. 77–85). Hong Kong: International Resources Management Institute.

Truong C. L. (n.d.). *Van Thuy Tu, Dinh Duc Thang, Duc Thang*. Unpublished manuscript, Phan Thiet (in Vietnamese).

Chapter 7
Conclusion: Errors and Insights

Kenneth Ruddle and Arif Satria

Abstract In addition to the erroneous assumption that tropical fisheries are 'open access' and not managed by pre-existing systems, and therefore require externally imposed management systems to protect resources from collapse and lift fishing communities out of poverty, the Western approach to fisheries 'development' and management suffers from several other basic flaws. These are that (1) pre-existing systems are as much, if not more, concerned with the community of fishers and their families and not just fisheries, and their principal role is ensuring community harmony and continuity; (2) pre-existing systems can involve multiple and overlapping rights that are flexible and adapted to changing needs and circumstances; (3) fisheries are just one component of a community resource assemblage with fisheries managed in their ecological context of being dependent on the good management of linked upstream ecosystems, and on risk management and ensuring balanced nutritional resources of the community; and (4) pre-existing systems are greatly affected by a constellation of interacting external pressures for change. If these cultural, ecological, economic, political and social context factors are not appreciated, any 'imposed management system' would likely fail from the outset to achieve its goals.

Keywords Fishing community management • Context factors • External pressures • Linked ecosystems • Risk

---

K. Ruddle (✉)
Asahigaoka-cho 7-22-511, Ashiya City, Hyogo-ken 659-0012, Japan
e-mail: mb5k-rddl@asahi-net.or.jp

A. Satria
Jalan Bambu Raya 30, Taman Yasmin Sektor 7, Bogor, Jawa Barat, Indonesia
e-mail: arifsatria@ipb.ac.id; arifsatria@gmail.com

K. Ruddle and A. Satria (eds.), *Managing Coastal and Inland Waters: Pre-existing Aquatic Management Systems in Southeast Asia*, DOI 10.1007/978-90-481-9555-8_7,

7.1 The Hegemonic Capitalistic-Industrial Model

The Modernization Theory, formulated in the 1950s and 1960s and inspired by the Marshall Plan for the post-World War II reconstruction of Europe, insisted that to successfully transplant social, economic and financial systems and values of industrial societies to poor tropical countries required that, because they were obstacles to development, their pre-existing economies and social systems be destroyed. Only by doing that could countries prosper and their people escape the trap of poverty.

As a consequence, tropical developing countries became enmeshed in a post-World War II system that is more nuanced than direct colonial administration, but nevertheless remains one in which the capitalistic-industrial model of the richest countries controls an uneven global resource exploitation that masquerades as 'development', 'growth' or 'progress'. However, in reality it is the underlying cause of the uneven impact of the interlinked global crises of environmental degradation, persistent poverty and food insecurity (Ruddle 2008). Dissident thinking that challenges the conformity with agendas set by political power brokers and the institutions that implement their policies is either filtered from the various funding processes, or co-opted and 'neutered', and thereby rendered harmless in a politically correct fashion. One such ordained solution, for example, has been to promote 'resilience thinking', based on a conformist approach of consensus, trivialization of problems and the blind implementation of pre-packaged policies (Homborg 2009).

7.2 The Mismanagement of Tropical Nearshore Fisheries

Within that context nearshore fisheries 'modernization' and management in the tropics have long been characterized by a Western scientific bias exacerbated by a lack of understanding or interest in local management systems, except where they hampered colonial administration (e.g., Ruddle 1995, 2007a), and an unwillingness or inability to try alternatives. As a consequence of development assistance, in the tropics as in temperate latitudes conservation of fish stocks became the main goal of transplanted Western fisheries management models, with fisheries policy and management based on a familiar bio-economic model.

The most socially pernicious impacts of this Western model derive from the erroneous assumption of the lack of prior local institutional arrangements among fishers to govern a fishery, and that fisheries are unregulated by local collective action. Thus the management model argues that to manage stock externalities institutional arrangements must be imposed on local fishing communities by some outside level of government. Such schemes are based on the falsehood that the institutional context of fisheries in the tropics is one of open access. Although this is simply not true for vast tracts of the world's nearshore waters, the lie was perfectly aligned with Western colonial and later assistance prescriptions for tropical developing countries.

Not uncommonly, fisheries biologists and social scientists who advise them have limited experience in the tropical milieux. Moreover, few fisheries scientists based in temperate latitude institutions sense any career advantage to working in the tropics. Not surprisingly, therefore, few appreciate the differences between the temperate zone industrial fisheries, with which they are familiar from their own training and research experience, and tropical nearshore fisheries. Consequently erroneous interpretations either are learned by or passed on to those who fund and make development policies and programs (Ruddle and Hickey 2008). For example, despite a huge amount of contrary evidence accumulated over the last three decades, it is still not well appreciated that opportunities to increase catches often are limited by exclusive rights systems. This is no small matter, because in many tropical areas marine tenure with associated rights limiting entry has been customary for centuries, with fisheries management based largely on such qualitative controls as limited access, closed seasons, areas and species, and a range of behavioral prohibitions which limited fishing pressure (Johannes 1978, 1982; Johannes and Hickey 2004; Hickey 2006). In contrast, an all too common and incorrect generalization is that the problem with fisheries lies in their open access nature (cf. Chapter 3, this volume).

7.3 Centralization Versus Decentralization

States exercise special claims to control common pool resources and manage the environment. In most countries fisheries management has usually been top-down, centralized, science-based and bureaucratic. Historically, that approach was basic to most development assistance programs. Criticisms of the failure of central management are commonplace. For example, fisheries management problems in Indonesia have been attributed to failures of the centralized fisheries system (Satria et al. 2002). In common with general criticisms, for Indonesia it was widely argued that environmental and resource concerns were subordinated to other development priorities, such that the central government was unwilling to address fisheries issues specifically. It also lacked human and financial resources to do so, should a compelling need have arisen. In addition, there was a limited financial and professional capacity to develop effective policy for resource and environmental management, as well as to monitor and enforce the implementation of any measures designed. Another important constraint has been resistance of resource users, owing to poor relationships between local communities and state authorities (see, for example, Bailey 1988; Bailey and Zerner 1992; Balland and Platteau 1996).

From the perspective taken in this volume, the critical aspect of that centralization policy is that, as in Indonesia, for example, it turned all nearshore marine waters into de facto open access, even though they were regulated de jure. This led to resource depletion, a main reason being that the high cost of enforcing centralized management made it impossible not to delegate to local people the responsibility for managing the great range of marine and coastal ecosystems. Meanwhile, many pre-existing fisheries management systems were undermined, and so could not bring their valuable social capital to bear on local management issues, owing to a lack of recognition and protection by central government. The de facto open access regime situation impelled 'free competition' among fishermen of different economic levels and ethnic and cultural backgrounds, among other distinctions. Resource depletion through overfishing and the destruction of mangroves and coral reefs was the inevitable result, which threatened the sustainability of fisheries, thereby leading to social conflicts.

Mostly conceived of as a failure, as evidenced by the collapse of fisheries managed according to this approach, it has been replaced by decentralization and approaches that emphasize participatory arrangements. But there are formidable stumbling blocks inherent in this model, too, and especially in the near-shore and small-scale fisheries of tropical countries (Ruddle and Hickey 2008). In the Philippines, decentralization was practiced from 1991, following enactment of the Local Government Code (Ebbers 2004), Section 16 of which gave a Local Government Unit or municipal government delegated authority for the management, conservation, development, protection, utilization, and disposition of all fish and fishery resources within their respective municipal waters. Further, the Fisheries Code of 1998 embodied the decentralization spirit, as demonstrated by the clarity of the roles, functions and responsibilities of the local government for coastal aquatic resources (Ebbers 2004). The Fisheries Code of 1998 promoted collaboration between government institutions and fishing communities in managing coastal fisheries resources (Ebbers 2004). In the Philippines the mechanisms to ensure community participation have been formalized, at least in law. In Indonesia decentralization was advanced by the Local Government Law of 1999, which empowered local governments to make decisions regarding local marine fisheries management. However, the way to institutionalize and establish decentralized marine fisheries management is still being considered. Nevertheless, the decentralization policy in Indonesia has had some positive impacts, particularly in the state recognition and strengthening of pre-existing fisheries management systems and the devolution of fisheries management to the community (Satria and Matsuda 2004a, b).

Decentralization is now regarded as the appropriate way to manage fisheries, by enabling local governments to control them. Ribot (2002) also justifies decentralization as a means of increasing the efficiency and equity of development activities and services delivery, since democratic decentralization is a promising means of institutionalizing and scaling up the popular participation that makes community-based natural resources management effective. However, Seddon (1999) argued that only if information flows between citizens and local government would the proximity of 'sub-national governments' to their constituents enable them to respond better to local needs and efficiently match public spending with private needs. On the other hand, the process of decentralization can itself enhance the opportunities for participation by placing more power and resources at a closer, more familiar and more easily influenced level of government. Hence, pre-existing fisheries management systems are potentially recognized, revitalized and developed. However, in addition to the potential positive impact of decentralization the process may lead to conflict, particularly when it involves the transfer of natural resources management and powers (Ribot 2002). But reality tempers such optimism.

7.4 Differing Perceptions of the New Western Decentralized Management Concepts

Community-based and co-management systems have been widely examined to replace the centralized system. The former refers to the form of management whereby a community controls the planning, implementation and evaluation of fisheries management. The latter is a practice by which fisheries management authorities and obligations are shared between government and the community (McCay 1995). Co-management has been particularly fashionable in recent years, and has been widely advocated regardless of its suitability to local social conditions, and commonly without regard for pre-existing systems of management.

The question of perceptions must be paramount in any evaluation of the suitability of the co-management approach, since at least national governments, donors and user communities are normally involved, regardless of the details of any local setting (Hara 2000). Each is likely to have different perceptions of the benefits, basis and hidden agendas in the co-management process, such that the inevitable conflict undermines programs or projects from the very beginning. Among the most pernicious of all these differing perceptions is that national governments and donors commonly assume that user communities are not organized, and that existing local institutions based on pre-existing systems and customary law are not suitable for use in a new management regime. As a consequence, it is usually assumed that the national government must organize and mobilize the community to participate effectively in the new management regime. Usually, the new institutions are created by government facilitators, nominally employing Western democratic principles and processes (Ruddle and Hickey 2008).

Common sense would suggest that the principal interest of fishers is not in the type of management system under which they operate so much as an improved household and community economic situation, and general social well-being. Although there is much theory, there is little practical evidence that co-management would contribute to this any more than the failed management regimes it is advocated to replace. It probably would not, since the principal general economic issue in fishing communities is not the condition of the fishery but of the narrow economic base of all rural communities.

Overcoming the weaknesses that constrain the ability of communities to manage themselves and their resource endowment in an integrated manner is a key to local development; not some elaborate plan devised from afar. Unfortunately, many of the Western-designed projects simply take too long, their design is too elaborate, they involve too many actors and levels, and there is far too much scope for sabotage. In contrast, many of the earlier studies on non-Western management systems (e.g., Johannes 1978, 1981, 1982) proposed using pre-existing local systems for a modern purpose in precisely those locations (e.g., Samoa, Vanuatu and Solomon Islands) where pre-existing systems remained either still functional or well remembered, as had been done effectively long ago in Japan (Ruddle 1987). Unfortunately, others, in the service of donors, latched onto these concepts and devised convoluted schemes to have 'locals' in many diverse parts of the tropical world want what they needed, whether they realized it or not. That was a serious misapplication of ideas.

7.5 Poorly Examined Basic Issues

In addition to the now familiar design principles, structure and content of authority, rights, rules, monitoring (etc.) and sanctions, the five studies in this volume have demonstrated some fundamental issues not usually examined in studies of pre-existing aquatic resources management systems. We summarize them here.

7.5.1 Managing Fishing Communities not Fisheries

The five cases in this volume demonstrate that each system described is as much, if not more, concerned with the community of fishers and their families as with fisheries per se. It can be argued that the principal role of the pre-existing management systems described in this book is ensuring community harmony and continuity, which commonly emphasizes importance of ancestors. This is clearly demonstrated for the cases from Batanes Province of the Philippines (Chapter 4) and Vietnam (Chapter 6), and to a lesser extent those from Indonesia (Chapter 2). It is also less apparent in the studies of Laos (Chapter 3) and Thailand (Chapter 4), which were designed to highlight other aspects of systems. Central to the systems described here for Indonesia, Batanes Province of the Philippines, and Vietnam is the role of sacred functions to achieving community harmony and continuity. It is less evident in the study on Laos (Chapter 3), but was examined by Tubtim (2006).

As explained in Chapter 2, under the awig–awig system of Lombok Island, Indonesia the coordinated authority resides in the mangku, an hereditary position the power of which is regarded as supernaturally rooted. Community members comply with the decisions of the mangku based on their belief in his sacred powers, which also represent community continuity through his inherited lineage. In his sacred role a mangku maintains the traditional community values regarding society and human relationships. His resource management role, which is based on his being a knowledgeable person with the secular ability to practice menjango, membanggar, and membuka, is also based on the sacred, since all of these secular activities/abilities are accompanied by religious ceremonies. Monitoring of rules regarding closed seasons and areal closure is done by lang–lang, the traditional coast 'police' appointed by the mangku laut, and therefore by extension deriving their authority from his sacred authority. Similarly, in Maluku Province, Indonesia, the sasi rules that govern the use of the petuanan are accompanied by religious ceremonies performed by a ritual practitioner, thus giving sasi also a sacred basis.

In Batanes Province, Philippines, the underlying function of the seasonal fishery is community management by ensuring the continuity of its values rooted in ancestor worship and associated symbolic behavior. Fishers make a 'vanua of the ancestors' based on sacred rites to ensure fishers' and community safety and a good catch. Authority is vested in a 'lead fisher', who makes the new season's first fishing trip and thereby inaugurates the vanua. He is a master fisher selected by the members of the vanua based on his experience and skill in fishing, and his good relationships within the community.

Symbolic authority consists of making the vanua and a ritual contract based on cooperation and conformity of the fishers, who ask the fish to appear and thereby ease the spirits of the community's ancestors. The principal rules governing the fishery reinforce this sacred nature of the fishery. In particular this can be seen for the set of rules that prescribe etiquette during fishing, when speaking of the fish, handling the catch, and distributing and eating fish. The object of this set of fish-related rules related to symbolic behavior is to maintain social harmony, order and cooperation in the community by all members showing respect for the Dorado fish. The only type of punishment described is social pressure to cease fishing temporarily on those perceived to be transgressing etiquette rules.

Management of the community and not just of the fishery and the role of the sacred is of paramount importance in Vietnam, where 'remembering the source from which one drinks' summarizes the focal importance of the ancestors and continuity of the community through interpersonal management that includes all fisheries stakeholders (Ruddle 2009). Although varying by locality, the veneration of deities and ancestors combined with the sacred obligations of mutual assistance remain the underlying and all-pervasive principles of a van chai. The linkage between shrine annual festivals and mutual assistance implies that mutual assistance is a sacred duty of van members, and thereby demonstrates the traditional moral authority of the van (Ruddle 1998, 2009; Ruddle and Luong 2009). The rituals performed at these festivals emphasize the importance of harmonious relationships among the various stakeholders in the fishery, and the vital importance of mutual assistance, both of which ensure the continued prosperity of the community and thereby venerate the ancestors. Mutual assistance obligations are specified in detail. The predominant design principles of a van chai are those reflecting and governing human relationships among the various stakeholders in the fishing community, in the context of mutual assistance and respect as governed by the precepts of the whale shrine. Despite years of turmoil, the core of the van chai system has proven remarkably resilient, undoubtedly because its salient characteristic is regulation of inter-relationships among fisheries stakeholders, within the framework of the strong moral authority of the community shrine, rather than regulation of fishing and the fishery per se.

7.5.2 *Complex, Flexible and Dual Rights Systems*

7.5.2.1 Complex Systems

As particularly well demonstrated for the Thai study (Chapter 5), pre-existing systems can involve multiple and overlapping rights that are nevertheless flexible and adapted to changing needs and circumstances. Rights revolve around a primary (i.e. a birthright) to access common property resources. Such an entitlement is meaningless if not accompanied by the right to exclude outsiders, or to negotiate access restricted by obligations, as is demonstrated in the cases from Maluku Province, Indonesia (Chapter 2), Laos (Chapter 3), and especially from Thailand (Chapter 5).

Whereas most of the commonly acknowledged rights in the literature emerge in these cases, straightforward assumptions regarding their discreetness either in form or operation are often revealed to be simplistic; rights are usually complex. The situation is varied and defies facile observation. Nevertheless, the fundamental rule of birthright operates in all cases except that of Batanes Province, Philippines, where any resident fisherman has the right to participate in a vanua and can select which he will belong to. Similarly the right of exclusion of outsiders or, better put, negotiated entry rights for them, is common, as in Maluku Province, Indonesia (Chapter 2), Laos (Chapter 3) and Thailand (Chapter 5).

In Maluku Province, Indonesia (Chapter 2), fisheries property rights are based on the concept of petuanan laut, a community-controlled exclusive territory, which convey a property right within an integrated estate system. In that territory community members have a birthright of hak makan. However, only the decedents of the original community founding group have the hak milik. Two basic rights are (1) hak makan ('the right to eat') and (2) hak milik ('the right of ownership'). Hak makan is compounded of the right of access and the right to use. Hak milik also contains the 'right of transfer'. The set of use rules governing a petuanan laut is known as sasi. The main ones pertain to access rights of outsiders. This is permitted via hak makan, but based on the concepts of transfer contained within hak milik. Sasi is accompanied by religious ceremonies performed by a ritual practitioner, which gives it a sacred basis. Authority is vested in the kewang, a traditional special committee headed by kepala kewang, who heads the village and leads the practice of sasi. There exist many detailed variations based on the type of resource area controlled, the belief system, the type of ritual leadership and the locations where the rites are performed.

In the Lower Songkhram River Basin (LSRB) of Northeast Thailand fishers operate under dual system comprised of the Fisheries Law of 1947, administered by the Department of Fisheries, and pre-existing village authority and rights. The latter recognizes that fishers have ownership of fishing rights areas and that they also have the right to exclude others from fishing within them. The result is a complex and multiple set of overlapping, complementary and conflicting individual, common and state property rights within a fishing ground (Khumsri et al. 2009). Communities recognize differing 'bundles' of de facto rights over fishing grounds, the ownership of which is restricted to those families, relatives or partners with traditionally established user rights over particular water bodies. The principal bundles of rights are (1) property rights as an authorized user, (2) property rights as a proprietor, and (3) property rights as an owner. As a consequence, all the best locations have long been owned. The basic features of these rights are that (1) owners can exclude others from their fishing ground, and (2) the rights can be sold,

7.5.2.2 Local Change in Rights Systems

The barrage fishery in LSRB of Northeast Thailand provides a detailed example of local change. As a result of both administrative change and the evolution of the rural economy since the 1950s, major changes have occurred in the barrage fishery. Formerly, grounds for it were owned by individuals as a private property. But from 1986 the barrage fishery was reclaimed by communities, and converted to a common property. Nowadays, possession of the de facto rights for the barrage fishery alternates between the community and individuals. Communities collectively agree to auction barrages and to decide access and use rules for them. Winning bidders are the authorized users, since they have only operational rights of access and withdrawal, and cannot establish management and exclusion rules. However, they can transfer and sell their harvesting rights, as when they sell them to small-scale fishers, and others may access the barrage areas for collecting wild foodstuffs, but not for fishing. Finally, after barrage operations cease the fishing grounds again become a common property open to the entire community (Khumsri et al. 2009).

7.5.2.3 Local Acceptance of Illegal Gear and the Conflict Between Local and State Rights

In the LSRB the consensus is that for both ecological and social reasons barrage fisheries have more negative than positive impacts. So most fishers and fisheries officers regard barrage fishing negatively, and agree with the Fisheries Law concerning its illegality. However, barrage fishing produces the highest fish yields of any large-scale gear used in the LSRB, and since this relates to the local communities' objective of maximizing revenue, rules are relaxed when applied to barrage fishing, and local DoF officers do not monitor compliance. As a result the barrage fishery is widespread and has gained increased political and economic importance under the auction system. This situation is tolerated by government, even though known to threaten the sustainability of fisheries resources (Khumsri 2008; Khumsri et al. 2009).

7.6 Fisheries Are Just One Component of a Community Resource Assemblage

The 'estate concept' in which fisheries are just one component has fundamental implications for management. Within the 'estate framework' fisheries are managed in their ecological context of being dependent of the good management of linked upstream ecosystems as well as risk management and ensuring balanced nutritional resources of the community (Ruddle 2011).

Although pre-existing systems of resource integrated resource management have long been widespread in tropical regions, they have not been well described for Southeast Asia. They are widespread in the South Pacific, and on high islands 'estates' are usually wedge-shaped, extending from a central watershed along lateral ridges into inshore marine waters. These are or were self-contained units that include a complete set of the resource areas and habitats required to provision the society that inhabited them. It remains a widespread integrated management strategy, as

Ruddle (1994) noted. Examples also occur in Africa (Manshard 1974; Ruttenberg 1980), and South America (Ruddle and Chesterfield 1977).

The awig–awig of Lombok Island, Indonesia (Chapter 2) provides a Southeast Asian example. It is based on the integrated estate concept known as sawen, a sophisticated human ecosystem concept that links ecological systems and resource assemblages from the upland forest though the coral reef (Satria 2007). Although each section of the longitudinal profile has its own management authority, with distinct roles and responsibilities for resource sustainability, their management roles were highly coordinated, which resulted in functional interdependence.

In the system described for Batanes Province, Philippines (Chapter 4) it is important to note that the maximization of the Dorado catch is not the object of this fishery. Rather, it is used as a currency with exchange partners to pay them for other economic services that they have performed for the fishermen (like farm labor during the Dorado fishing season, for example). So in this sense in addition to being a system for community management, it could be envisaged as an 'integrated estate system of the mind'.

7.7 Change

Among the common major external forces causing change are the legacy of colonialism, contemporary government policy and legal change, and the replacement of traditional local authority. Demographic change, 'modernization', economic development, commercialization and commoditization of resources, technological change, donor conditionalities, and national policies for economic sectors other than fisheries collectively have generally resulted in changes in the perceptions of fishing communities regarding the value of aquatic resources. Often external factors are internalized by villages elites, which can lead to the breakdown or weakening of pre-existing systems of management, all of which arose and developed within a specific social context to meet particular needs. Communities are not immune to the pressures that drive larger polities and commercial elites. Expanded markets introduce temptations for individuals to profit at the expense of the community equity and allocation of resources and thus undermine the moral authority of systems (as in the Kei Islands of Indonesia, Chapter 2). As a consequence, participants in pre-existing systems "... cannot be assumed a priori as being inherently benign resource-conservational and socially equitable actors. Hence any policy and program decisions about the present-day and future usefulness of local management systems must be based on a clearheaded and realistic evaluation of the moral authority, motives, interests and cultural conceptions that underpin and drive them" (Ruddle 1993:2).

Many of the earlier studies on non-Western management systems proposed using pre-existing local systems for a modern purpose in precisely those locations (e.g., Samoa and Vanuatu) where pre-existing systems remained either still functional or well remembered (Johannes 1998, 2002). That approach was applied with the now clearly visible excellent results, meriting reinforcement and wider application. To do that demands a radically different approach to fisheries management (Ruddle 2007b), which recognizes (1) that the underlying characteristics of nearshore fisheries in tropical countries are vastly different from those for which the conventional Western approaches were developed; (2) that the various Western approaches to managing fisheries have not been successful in tropical nearshore fisheries; and (3) that there exist in many tropical developing countries pre-existing systems that provide proven alternative approaches to management and blue prints for new systems, since they are already pre-adapted to the characteristics of tropical nearshore fisheries and cultural milieux.

However, in many other places, including much of Southeast Asia, economic, political and related change triggers an alteration of management and property rights regimes (e.g., Ruddle 1993). In Vietnamese fishing communities, for example, the van chai was formed centuries ago in areas with little population and small and local markets for aquatic products, where there was little pressure on aquatic resources, and where fishing boats and gear were just those needed to harvest for local consumption. Other than for administration, such fishing communities were little connected with the larger national social organization. In human terms the van chai was based on 'neighborly affection', in which behavior and social standards were rooted in Confucianism, which remains fundamental in Vietnamese society. Satisfaction of the spiritual needs of its fishing community was a main function of a van chai.

But things can change fast, as during the post-colonial era, from 1954 to 1975, when conditions differed in the former Democratic Republic of Vietnam, in the north, and the Republic of Vietnam, in the south. In the former, coastal waters belonged to the state and governance of local waters was by the commune, the lowest level in the government structure. Fisheries production was collectivized. In the south, ownership rights remained unchanged; fishers leased rights from and were taxed by village governments. With re-unification, in 1975, the government initially extended nationwide the system prevailing in the north. Previous administrative organization and rights were abandoned, and the management of local fisheries according to national laws was handed to the provinces and the communes. During the period of cooperative development, the national government paid no attention to the van chai, so they declined. However, after the near universal collapse of fishing cooperatives, the national government is beginning to appreciate the potential of van chai as a vehicle for local fisheries management (Ruddle and Tuong 2009).

Despite those decades of turmoil and change, in many areas the systems have survived. It is noteworthy that because the salient characteristic of traditional management systems in Vietnam is regulation of inter-relationships among fisheries stakeholders, within the framework of the strong moral authority of the community shrine, rather than governance of fishing and the fishery per se, the core of the system has proven remarkably resilient. Nevertheless, the van chai is now strongly influenced by external forces, such that its autonomous identity has disappeared. As a consequence local social norms and standards have changed radically, undermining the social mechanism based on pre-existing or customary rules for van chai operations.

Further, the pressure on aquatic resources is now intense, with a greatly enlarged market that now includes all of Vietnam plus a large international demand. The means of production are now so intense that the resource has been greatly depleted. As a result of such massive changes in context, it is an illusion to consider restructuring the van chai system exactly in its pre-existing form. However, this does not preclude using the underlying principles embodied in the philosophy and pre-existing rules of the older van chai, to empower all fishing communities to participate in sustainably managing aquatic resources. The challenge is to create an appropriate legal framework for doing that. This is the situation throughout Southeast Asia.

References

Bailey, C. (1988). The political economy of marine fisheries development in Indonesia. *Indonesia, 46*, 25–38.

Bailey, C., & Zerner, C. (1992). Community-based fisheries management institutions in Indonesia. *Maritime Anthropological Studies, 5*(1), 1–17.

Balland, J., & Platteau, J. (1996). *Halting degradation of natural resources: Is there a role for rural communities?* New York: Oxford University Press.

Ebbers, T. (2004). Making the Law Work: Decentralization and right-based fisheries in the Philippines. *Fish for the People, 2*(2), 2–10.

Hara, M. (2000, November). *Fisheries co-management; what and whose agenda?* Paper presented at the WARFSA/WaterNet Symposium: Sustainable Use of Water Resources, Maputo.

Hickey, F. R. (2006). Traditional marine resource management in Vanuatu: Acknowledging, supporting and strengthening indigenous management systems. *SPC Traditional Marine Resource Management and Knowledge Information Bulletin, 20*, 11–23.

Homborg, A. (2009). Zero-sum world: Challenges in conceptualizing environmental load displacement and ecologically unequal exchange in the world system. *International Journal of Comparative Sociology, 50*(3–4), 237–262.

Johannes, R. E. (1978). Traditional marine conservation methods in Oceania and their demise. *Annual Review of Ecology and Systematics, 9*, 349–64.

Johannes, R. E. (1981). *Words of the lagoon: Fishing and marine lore in the Palau District of Micronesia*. Berkeley: University of California Press.

Johannes, R. E. (1982). Traditional conservation methods and protected marine areas in Oceania. *Ambio, 11*(5), 258–261.

Johannes, R. E. (1998). Government-supported, village-based management of marine resources in Vanuatu. *Ocean and Coastal Management, 40*, 165–186.

Johannes, R. E. (2002). The renaissance of community-based marine resource management in oceania. *Annual Review of Ecology and Systematics, 33*, 317–340.

Johannes, R. E., & Hickey, F. R. (2004). *Evolution of village-based marine resource management in Vanuatu between 1993 and 2001*. Coastal Region and Small Island Papers 15. Paris: UNESCO.

Khumsri, M. (2008). *Co-Management institutions for sustainable inland fisheries management: A case study of the Lower Songkhram River Basin, Thailand*. Dissertation, Asian Institute of Technology.

Khumsri, M., Ruddle, K., & Shivakoti, G. P. (2009). Rights and conflicts in the management of fisheries in the Lower Songkhram River Basin, Northeast Thailand. *Environmental Management, 43*, 557–570.

Manshard, W. (1974). *Tropical agriculture: A geographical introduction and appraisal*. London: Longman.

McCay, B. J. (1995). Common and private concerns. *Advances in Human Ecology, 4*, 90–116.

Ribot, J. C. (2002). *Democratic decentralization of natural resources: Institutionalizing popular participation*. Washington, DC: World Resources Institute.

Ruddle, K. (1987). *Administration and conflict management in Japanese coastal fisheries*, FAO Fisheries Reports 273. Rome: FAO.

Ruddle, K. (1993). External forces and change in traditional community-based fishery management systems in the Asia-Pacific region. *Maritime Anthropological Studies, 6*(1–2), 1–37.

Ruddle, K. (1994). *A guide to the literature on traditional community-based fishery management in the Asia-Pacific tropics*. Fisheries Circular No. 869, FIPP/C869. Rome: FAO.

Ruddle, K. (1995). The role of validated local knowledge in the restoration of fisheries property rights: the example of the New Zealand Maori. In S. Hanna & M. Munasinghe (Eds.), *Property rights in a social and ecological context: Vol. 2 case studies and design applications* (pp. 111–120). Stockholm/Washington, DC: The Beijer International Institute of Ecological Economics and The World Bank.

Ruddle, K. (1998). Traditional community-based coastal marine fisheries management in Viet Nam. *Ocean and Coastal Management, 40*(3), 1–22.

Ruddle, K. (2007a). Wronging rights and righting wrongs. In W. Taylor, M. Schechter, & L. Wolfson (Eds.), *Globalization: Effects on fisheries resources* (pp. 215–228). Cambridge: Cambridge University Press.

Ruddle, K. (2007b). Fishing rights: misconceptions, outright prejudice. *Samudra The Triannual Journal of the International Collective in Support of Fishworkers, Report No* 48: 4–9.

Ruddle, K. (2008). Reconsidering the contribution of fisheries to society and Millennium Development Goals. In K. Tsukamoto, T. Kawamura, T. Takeuchi, T. D. Beard, & M. J. Kaiser (Eds.), *Fisheries for global welfare and environment, 5th World Fisheries Congress, 2008* (pp. 199–411). Tokyo: Terra Scientific Publishing Co.

Ruddle, K. (2009). Remembering the Source. *Samudra The Triannual Journal of the International Collective in Support of Fishworkers Report, 54*, 10–13.

Ruddle, K. (2011). Coastal-marine Resource Use in Human Ecological Context: the Scale and Modes of Integration. In: V. Christensen, & J. Maclean (Eds.), *Ecosystem approaches to fisheries: A global perspective,* Cambridge: Cambridge University Press (in press).

Ruddle, K., & Chesterfield, R. A. (1977). *Education for traditional food procurement in the Orinoco Delta. Iberoamericana No. 53*. Berkeley/Los Angeles: University of California Press.

Ruddle, K., & Hickey, F. R. (2008). Accounting for the mismanagement of tropical nearshore fisheries. *Environment, Development and Sustainability, 10*(5), 565–589.

Ruddle, K., & Luong, T. S. (2009). The importance of *van chai* in fisheries village development and management in Binh Thuan Province. In K. Ruddle & P. L. Tuong (Eds.), *The van chai of Vietnam: Managing nearshore fisheries and fishing communities* (pp. 47–65). Hong Kong: International Resources Management Institute.

Ruddle, K., & Tuong, P. L. (2009). *The van chai of Vietnam: Managing nearshore fisheries and fishing communities*. Hong Kong: International Resources Management Institute.

Ruttenberg, H. (1980). *Farming systems in the tropics*. Oxford: Clarendon House.

Satria, A. (2007). Sawen: institution, local knowledge and myths in fisheries management in North Lombok, Indonesia. In N. Haggan, B. Neis, & I. G. Baird (Eds.), *Fishers' knowledge in fisheries science and fisheries management* (pp. 199–220). Paris: UNESCO.

Satria, A., & Matsuda, Y. (2004b). Decentralization policy: An opportunity for strengthening fisheries management system. *Journal of Environment and Development, 13*(2), 179–196.

Satria, A., Umbari, A., Fauzi, A., Purbayanto, A., Sutarto, E., Muchsin, I., et al. (2002). *Menuju Desentralisasi Kelautan*. Jakarta: Cidesindo (in Indonesian).

Seddon, J. (1999). *Participation, civil society, and decentralization. Decentralization briefing notes*. Washington, DC: World Bank Institute.

Tubtim, N. (2006). Exclusion, accommodation and community-based natural resource management: Legitimizing the enclosure of a community fishery in southern Laos. In S. Tyler (Ed.), *Communities, livelihoods, and natural resources: Action research and policy change in Asia* (pp. 1–17). Ottawa: IDRC.

N/A →

Author Index

Subject Index

D

Breinigsville, PA USA
19 September 2010
245653BV00003B/36/P